CRISC™ REVIEW QUESTIONS, ANSWERS & EXPLANATIONS MANUAL 2015

CRISC™ Certified in Risk and Information Systems Control™

An ISACA® Certification

ISACA®

With more than 115,000 constituents in 180 countries, ISACA (*www.isaca.org*) helps business and IT leaders build trust in, and value from, information and information systems. Established in 1969, ISACA is the trusted source of knowledge, standards, networking, and career development for information systems audit, assurance, security, risk, privacy and governance professionals. ISACA offers the Cybersecurity Nexus™, a comprehensive set of resources for cybersecurity professionals, and COBIT®, a business framework that helps enterprises govern and manage their information and technology. ISACA also advances and validates business-critical skills and knowledge through the globally respected Certified Information Systems Auditor® (CISA®), Certified Information Security Manager® (CISM®), Certified in the Governance of Enterprise IT® (CGEIT®) and Certified in Risk and Information Systems Control™ (CRISC™) credentials. The association has more than 200 chapters worldwide.

Disclaimer

ISACA has designed and created *CRISC™ Review Questions, Answers & Explanations Manual 2015* primarily as an educational resource to assist individuals preparing to take the CRISC certification exam. It was produced independently from the CRISC exam and the CRISC Certification Committee, which has had no responsibility for its content. Copies of past exams are not released to the public and were not made available to ISACA for preparation of this publication. ISACA makes no representations or warranties whatsoever with regard to these or other ISACA publications assuring candidates' passage of the CRISC exam.

Reservation of Rights

ISACA
3701 Algonquin Road, Suite 1010
Rolling Meadows, IL 60008 USA
Phone: +1.847.253.1545
Fax: +1.847.253.1443
Email: *info@isaca.org*
Web site: *www.isaca.org*

Participate in the ISACA Knowledge Center: *www.isaca.org/knowledge-center*
Follow ISACA on Twitter: *https://twitter.com/ISACANews*
Join ISACA on LinkedIn: ISACA (Official), *http://linkd.in/ISACAOfficial*
Like ISACA on Facebook: *www.facebook.com/ISACAHQ*

ISBN 978-1-60420-591-6
CRISC™ Review Questions, Answers & Explanations Manual 2015
Printed in the United States of America

PREFACE

ISACA is pleased to offer the 400 questions in this *CRISC™ Review Questions, Answers & Explanations Manual 2015*. The purpose of this publication is to provide the CRISC candidate with sample questions and testing topics to help prepare and study for the CRISC exam.

The material in this manual consists of 400 multiple-choice questions, answers and explanations intended to introduce CRISC candidates to the types of questions that may appear on the CRISC exam. They are not actual questions from the exam. All of these items appeared the *CRISC Review Questions, Answers & Explanations Manual 2013*, the *CRISC Review Questions, Answers & Explanations Manual 2013 Supplement* and the *CRISC Review Questions, Answers & Explanations Manual 2014 Supplement*. The 400 questions are sorted by CRISC job practice domains. Additionally, 150 questions have been extracted to provide a sample exam with questions in the same proportion as the current CRISC job practice. The candidate also may want to obtain a copy of the *CRISC™ Review Manual 2015*, which provides the foundational knowledge of a CRISC, and the *CRISC™ Review Questions, Answers & Explanations Manual 2015 Supplement*, which consists of 100 new multiple-choice study questions. The candidate also may wish to obtain the questions in a web-based format in the CRISC™ Review Questions, Answers & Explanations Database - 12 Month Subscription.

Some of the questions are presented in scenarios. Scenarios are mini-case studies that describe a situation or an enterprise and require candidates to answer one or more questions based on the information provided. A scenario can focus on one or more domains. The CRISC exam may include scenario questions.

ISACA has produced this publication as an educational resource to assist individuals preparing to take the CRISC exam. It was produced independently from the CRISC Certification Committee, which has no responsibility for its content. Copies of past exams are not released to the public and are not made available to candidates. ISACA makes no representations or warranties whatsoever with regard to these or other ISACA or IT Governance Institute publications assuring candidates' passage of the CRISC exam.

ISACA wishes you success with the CRISC exam and welcomes your comments and suggestions on the use and coverage of this manual. Once you have completed your exam, please take a moment to complete the online evaluation that corresponds to this publication (*www.isaca.org/studyaidsevaluation*). Your observations will be invaluable as new questions, answers and explanations are prepared.

ACKNOWLEDGMENTS

This *CRISC™ Review Questions, Answers & Explanations Manual 2015* is the result of the collective efforts of many volunteers. ISACA members from throughout the world participated, generously offering their talent and expertise. This international team exhibited a spirit and selflessness that has become the hallmark of contributors to this valuable manual. Their participation and insight are truly appreciated.

NEW—CRISC JOB PRACTICE

BEGINNING IN 2015, THE CRISC EXAM WILL TEST THE NEW CRISC JOB PRACTICE.

An international job practice analysis is conducted at least every five years or sooner to maintain the validity of the CRISC certification program. A new job practice forms the basis of the CRISC exam beginning in June 2015.

The primary focus of the job practice is the current tasks performed and the knowledge used by CRISCs. By gathering evidence of the current work practice of CRISCs, ISACA is able to ensure that the CRISC program continues to meet the high standards for the certification of professionals throughout the world.

The findings of the CRISC job practice analysis are carefully considered and directly influence the development of new test specifications to ensure that the CRISC exam reflects the most current best practices.

The new 2015 job practice reflects the areas of study to be tested and is compared below to the previous job practice.

Previous CRISC Job Practice	New 2015 CRISC Job Practice
Domain 1: Risk Identification, Assessment and Evaluation (31%) Domain 2: Risk Response (17%) Domain 3: Risk Monitoring (17%) Domain 4: Information Systems Control Design and Implementation (17%) Domain 5: Information Systems Control Monitoring and Maintenance (18%)	Domain 1: IT Risk Identification (27%) Domain 2: IT Risk Assessment (28%) Domain 3: Risk Response and Mitigation (23%) Domain 4: Risk and Control Monitoring and Reporting (22%)

Page intentionally left blank

TABLE OF CONTENTS

Page intentionally left blank

INTRODUCTION

The *CRISC™ Review Questions, Answers & Explanations Manual 2015* has been developed to assist the CRISC candidate in studying and preparing for the CRISC exam. As you use this publication to prepare for the exam, please note that the exam covers a broad spectrum of IS control solutions and how they relate to business and IT risk management issues. Do not assume that reading and working the questions in this manual will fully prepare you for the exam. Since exam questions often relate to practical experience, CRISC candidates are advised to refer to their own experience and to other publications and frameworks referred to in the *CRISC™ Review Manual 2015*, such as the COBIT 5 framework and *COBIT 5 for Risk*. These additional references are excellent sources of further detailed information and clarification. It is suggested that candidates evaluate the domains in which they feel weak or require a further understanding and then study accordingly.

DOCUMENT STRUCTURE

This manual consists of 400 sample multiple-choice questions, answers and explanations. These questions are provided in two formats:
1. Questions Sorted by Domain
2. Sample Exam

1. Questions Sorted by Domain

Questions, answers and explanations are provided (sorted) by CRISC job practice domains. This allows the CRISC candidate to refer to specific questions to evaluate comprehension of the topics covered within each domain. These questions are representative of CRISC questions, although they are not actual exam items. They are provided to assist the CRISC candidate in understanding the material in the *CRISC™ Review Manual 2015* and to depict the type of question format typically found on the CRISC exam. The numbers of questions, answers and explanations provided in the four domain chapters in this publication provide the CRISC candidate with a maximum number of study questions.

Scenarios

Some of the questions are presented in scenarios. Scenarios are mini-case studies that describe a situation or an enterprise and require candidates to answer one or more questions based on the information provided. A scenario can focus on one or more domains. The CRISC exam may include scenario questions.

2. Sample Exam

A random sample exam of 150 of the questions is also provided in this manual. **This exam is organized according to the domain percentages specified in the CRISC job practice and used on the CRISC exam:**

Domain 1—IT Risk Identification .... 27 percent
Domain 2—IT Risk Assessment .... 28 percent
Domain 3—Risk Response and Mitigation .... 23 percent
Domain 4—Risk and Control Monitoring and Reporting .... 22 percent

Candidates are urged to use this sample exam and the answer sheets provided in this publication to simulate an actual exam. Many candidates use this sample exam as a pretest to determine their specific strengths or weaknesses, or as a final test to determine their readiness to sit for the exam. Sample exam answer sheets have been provided for both uses, and an answer/reference key is included. These sample exam questions have been cross-referenced to the questions, answers and explanations by domain so it is convenient to refer back to the explanations of the correct answers. This publication is ideal to use in conjunction with the *CRISC™ Review Questions, Answers & Explanation Manual 2015 Supplement*.

It should be noted that the *CRISC™ Review Questions, Answers & Explanations Manual 2015* has been developed to assist the CRISC candidate in studying and preparing for the CRISC exam. As you use this publication for the exam, please recognize that individual perceptions and experiences may not reflect the more global position or circumstance. Since the CRISC exam and manuals are written from a global perspective, the candidate will be required to be somewhat flexible when reading about a condition that may be contrary to the candidate's experience. It should be noted that actual CRISC exam questions are written by experienced IS risk and control practitioners from around the world. Each question on the exam is reviewed by ISACA's CRISC Test Enhancement Subcommittee and ISACA's CRISC Certification Committee, both of which consist of international members. This geographic representation ensures that all test questions will be understood equally in each country and language.

Also, please note that this publication has been written using standard American English.

TYPES OF QUESTIONS ON THE CRISC EXAM

CRISC exam questions are developed with the intent of measuring and testing practical knowledge and applying general concepts and standards. As previously mentioned, all questions are presented in a multiple choice format and are designed for one best answer.

The candidate is cautioned to read each question carefully. Many times, a CRISC exam question will require the candidate to choose the appropriate answer that is **MOST** likely or **BEST**. Other times, a candidate may be asked to choose a practice or procedure that would be performed **FIRST** related to the other choices. In every case, the candidate is required to read the question carefully, eliminate known wrong choices and then make the best choice possible. Knowing that these types of questions are asked and how to study to answer them will go a long way toward answering them correctly.

Each CRISC question has a stem (question) and four choices (answers). The candidate is asked to choose the best answer from the choices. The stem may be in the form of a question or an incomplete statement. In some instances, a scenario or description problem may also be included. These questions normally include a description of a situation and require the candidate to answer two or more questions based on the information provided.

> **Note:** ISACA review manuals are living documents. As technology advances, ISACA manuals will be updated to reflect such advances. Further updates or corrections to this document before the date of the exam may be viewed at *www.isaca.org/studyaidupdates.*

Any suggestions to enhance the materials covered herein, or reference materials, should be submitted online at *www.isaca.org/studyaidsevaluation.*

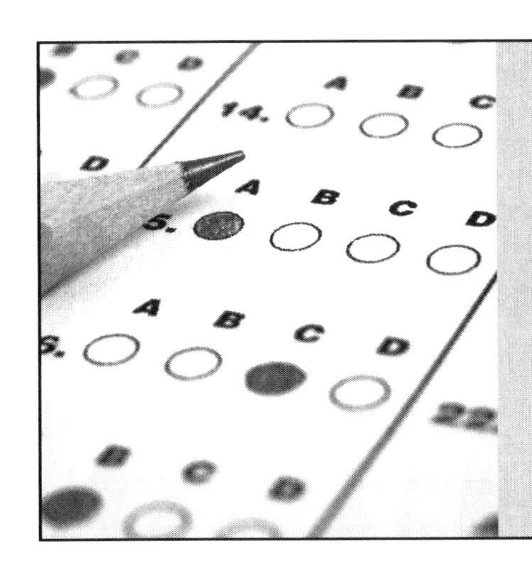

PRETEST

If you wish to take a pretest to determine strengths and weaknesses, the Sample Exam begins on page 181 and the pretest answer sheet begins on page 205. You can score your pretest with the Sample Exam Answer and Reference Key on page 203.

Page intentionally left blank

QUESTIONS, ANSWERS AND EXPLANATIONS BY DOMAIN

DOMAIN 1—IT RISK IDENTIFICATION (27%)

R1-1 Which of the following is **MOST** important to determine when defining risk management strategies?

A. Risk assessment criteria
B. IT architecture complexity
C. An enterprise disaster recovery plan (DRP)
D. Organizational objectives

D is the correct answer.

Justification:
A. Risk assessment criteria become part of this framework, but only after proper analysis.
B. IT architecture complexity is more directly related to assessing risk than defining strategies.
C. An enterprise disaster recovery plan (DRP) is more directly related to assessing risk than defining strategies.
D. While defining risk management strategies, the risk practitioner needs to analyze the organization's objectives and risk tolerance and define a risk management framework based on this analysis. Some organizations may accept known risk, while others may invest in and apply mitigating controls to reduce risk.

R1-2 Which of the following is the **MOST** important information to include in a risk management strategic plan?

A. Risk management staffing requirements
B. The risk management mission statement
C. Risk mitigation investment plans
D. The current state and desired future state

D is the correct answer.

Justification:
A. Risk management staffing requirements are generally driven by a robust understanding of the current and desired future state.
B. The risk management mission statement is important, but is not an actionable part of a risk management strategic plan.
C. Risk mitigation investment plans are generally driven by a robust understanding of the current and desired future state.
D. It is most important to paint a vision for the future and then draw a road map from the starting point; therefore, this requires that the current state and desired future state be fully understood.

R1-3 Information that is no longer required to support the main purpose of the business from an information security perspective should be:

A. analyzed under the retention policy.
B. protected under the information classification policy.
C. analyzed under the backup policy.
D. protected under the business impact analysis (BIA).

A is the correct answer.

Justification:

A. Information that is no longer required should be analyzed under the retention policy to determine whether the organization is required to maintain the data for business, legal or regulatory reasons. Keeping data that are no longer required unnecessarily consumes resources; may be in breach of legal and regulatory obligations regarding retention of data; and, in the case of sensitive personal information, can increase the risk of data compromise.

B. The information classification policy should specify retention and destruction of information that is no longer of value to the core business, as applicable.

C. The backup policy is generally based on recovery point objectives (RPOs). The information classification policy should specify retention and destruction of backup media.

D. A business impact analysis (BIA) can help determine that this information does not support the main objective of the business, but does not indicate the action to take.

R1-4 An enterprise has outsourced the majority of its IT department to a third party whose servers are in a foreign country. Which of the following is the **MOST** critical security consideration?

A. A security breach notification may get delayed due to the time difference.
B. Additional network intrusion detection sensors should be installed, resulting in additional cost.
C. The enterprise could be unable to monitor compliance with its internal security and privacy guidelines.
D. Laws and regulations of the country of origin may not be enforceable in the foreign country.

D is the correct answer.

Justification:

A. Security breach notification is not a problem. Time difference does not play a role in a 24/7 environment. Pagers, cell phones, etc., are usually available to communicate a notification.

B. The need for additional network intrusion sensors is a manageable problem that requires additional funding, but can be addressed.

C. Outsourcing does not remove the enterprise's responsibility regarding internal requirements.

D. Laws and regulations of the country of origin may not be enforceable in the foreign country. Conversely, the laws and regulations of the foreign outsourcer may also impact the enterprise. A potential violation of local laws applicable to the enterprise or the vendor may not be recognized or rectified due to the lack of knowledge of the local laws that are applicable and the inability to enforce those laws.

R1-5 An enterprise recently developed a breakthrough technology that could provide a significant competitive edge. Which of the following **FIRST** governs how this information is to be protected from within the enterprise?

A. The data classification policy
B. The acceptable use policy
C. Encryption standards
D. The access control policy

A is the correct answer.

Justification:

A. A data classification policy describes the data classification categories; level of protection to be provided for each category of data; and roles and responsibilities of potential users, including data owners.
B. An acceptable use policy is oriented more toward the end user and, therefore, does not specifically address which controls should be in place to adequately protect information.
C. Mandated levels of protection, as defined by the data classification policy, should drive which levels of encryption will be in place.
D. Mandated levels of protection, as defined by the data classification policy, should drive which access controls will be in place.

R1-6 Malware has been detected that redirects users' computers to web sites crafted specifically for the purpose of fraud. The malware changes domain name system (DNS) server settings, redirecting users to sites under the hackers' control. This scenario **BEST** describes a:

A. man-in-the-middle (MITM) attack.
B. phishing attack.
C. pharming attack.
D. social engineering attack.

C is the correct answer.

Justification:

A. In a man-in-the-middle (MITM) attack, the attacker intercepts the communication stream between two parts of the victim system and then replaces the traffic between the two components with the intruder's own, eventually assuming control of the communication.
B. A phishing attack is a type of email attack that attempts to convince a user that the originator is genuine, but with the intention of obtaining information for use in social engineering.
C. A pharming attack is an MITM attack that changes the pointers on a domain name system (DNS) server and redirects a user's session to a masquerading web site.
D. A social engineering attack deceives users or administrators at the target site into revealing confidential or sensitive information. They can be executed person-to-person, over the telephone or via email.

R1-7 What is the **MOST** effective method to evaluate the potential impact of legal, regulatory and contractual requirements on business objectives?

A. A compliance-oriented gap analysis
B. Interviews with business process stakeholders
C. A mapping of compliance requirements to policies and procedures
D. A compliance-oriented business impact analysis (BIA)

D is the correct answer.

Justification:

A. A gap analysis will only identify the gaps in compliance to current requirements and will not identify impacts to business objectives or activities.
B. Interviews with key business process stakeholders will identify business objectives, but will not necessarily account for the compliance requirements that must be met.
C. Mapping requirements to policies and procedures will identify how compliance is being achieved, but will not identify business impact.
D. A compliance-oriented business impact analysis (BIA) will identify all of the compliance requirements to which the enterprise has to align and their impacts on business objectives and activities.

R1-8 Which of the following is the **BEST** way to ensure that an accurate risk register is maintained over time?

A. Monitor key risk indicators (KRIs), and record the findings in the risk register.
B. Publish the risk register centrally with workflow features that periodically poll risk assessors.
C. Distribute the risk register to business process owners for review and updating.
D. Utilize audit personnel to perform regular audits and to maintain the risk register.

B is the correct answer.

Justification:

A. Monitoring key risk indicators (KRIs) will only provide insights to known and identified risk and will not account for risk that has yet to be identified.
B. Centrally publishing the risk register and enabling periodic polling of risk assessors through workflow features will ensure accuracy of content. A knowledge management platform with workflow and polling features will automate the process of maintaining the risk register.
C. Business process owners typically cannot effectively identify risk to their business processes. They may not have the ability to be unbiased in their review and may not have the appropriate skills or tools to effectively evaluate risk.
D. Audit personnel may not have the appropriate business knowledge or training in risk assessment to appropriately identify risk. Regular audits of business processes can also be a hindrance to business activities and will most likely not be allowed by business leadership.

R1-9 Shortly after performing the annual review and revision of corporate policies, a risk practitioner becomes aware that a new law may affect security requirements for the human resources system. The risk practitioner should:

A. analyze what systems and technology-related processes may be impacted.
B. ensure necessary adjustments are implemented during the next review cycle.
C. initiate an ad hoc revision of the corporate policy.
D. notify the system custodian to implement changes.

A is the correct answer.

Justification:

A. Assessing what systems and technology-related processes may be impacted is the best course of action. The analysis must also determine whether existing controls already address the new requirements.
B. Ensuring necessary adjustments are implemented during the next review cycle is not the best answer, particularly in cases where the law does affect the enterprise. While an annual review cycle may be sufficient in general, significant changes in the internal or external environment should trigger an ad hoc reassessment.
C. Initiating an ad hoc amendment to the corporate policy may be a rash and unnecessary action.
D. Notifying the system custodian to implement changes is inappropriate. Changes to the system should be implemented only after approval by the process owner.

R1-10 Which of the following is the **PRIMARY** objective of a risk management program?

A. Maintain residual risk at an acceptable level
B. Implement preventive controls for every threat
C. Remove all inherent risk
D. Reduce inherent risk to zero

A is the correct answer.

Justification:

A. Ensuring that all residual risk is maintained at a level acceptable to the business is the objective of a risk management program.
B. The objective of a risk management program is not to implement controls for every threat.
C. A risk management program is not intended to remove every identified risk.
D. Inherent risk—the risk level of an activity, business process or entity without taking into account the actions that management has taken or may take—is always greater than zero.

R1-11 Assessing information systems risk is **BEST** achieved by:

A. using the enterprise's past actual loss experience to determine current exposure.
B. reviewing published loss statistics from comparable organizations.
C. evaluating threats associated with existing information systems assets and information systems projects.
D. reviewing information systems control weaknesses identified in audit reports.

C is the correct answer.

Justification:
A. Past actual loss experience is a potentially useful input to the risk assessment process, but it does not address realistic risk scenarios that have not occurred in the past.
B. Published loss statistics from comparable organizations is a potentially useful input to the risk assessment process, but does not address enterprise specific risk scenarios or those that have not occurred in the past.
C. To assess IT risk, threats and vulnerabilities need to be evaluated using qualitative or quantitative risk assessment approaches.
D. Control weaknesses and other vulnerabilities are an important input to the risk assessment process, but by themselves are not useful.

R1-12 Which of the following is the **MOST** important requirement for setting up an information security infrastructure for a new system?

A. Performing a business impact analysis (BIA)
B. Considering personal devices as part of the security policy
C. Basing the information security infrastructure on a risk assessment
D. Initiating IT security training and familiarization

C is the correct answer.

Justification:
A. Typically, a business impact analysis (BIA) is carried out to prioritize business processes as part of a business continuity plan (BCP).
B. While considering personal devices as part of the security policy may be a consideration, it is not the most important requirement.
C. The information security infrastructure should be based on a risk assessment.
D. Initiating IT security training may not be important for the purpose of the information security infrastructure.

R1-13 The **PRIMARY** concern of a risk practitioner reviewing a formal data retention policy is:

A. storage availability.
B. applicable organizational standards.
C. generally accepted industry best practices.
D. business requirements.

D is the correct answer.

Justification:
A. Storage is irrelevant because whatever is needed must be provided.
B. Applicable organizational standards support the policy, but do not dictate it.
C. Best practices may be a useful guide, but not a primary concern.
D. The primary concern is business requirements.

R1-14 Which of the following areas is **MOST** susceptible to the introduction of an information-security-related vulnerability?

A. Tape backup management
B. Database management
C. Configuration management
D. Incident response management

C is the correct answer.

Justification:

A. Tape backup management is generally less susceptible to misconfiguration issues than configuration management.
B. Database management is generally less susceptible to misconfiguration issues than configuration management.
C. Configuration management provides the greatest likelihood of information security weaknesses through misconfiguration and failure to update operating system (OS) code correctly and on a timely basis.
D. Incident response management is generally less susceptible to misconfiguration issues than configuration management.

R1-15 Which of the following is the **MOST** important reason for conducting security awareness programs throughout an enterprise?

A. Reducing the risk of a social engineering attack
B. Training personnel in security incident response
C. Informing business units about the security strategy
D. Maintaining evidence of training records to ensure compliance

A is the correct answer.

Justification:

A. Social engineering is the act of manipulating people into divulging confidential information or performing actions that allow an unauthorized individual to get access to sensitive information and/or systems. People are often considered the weakest link in security implementations and security awareness would help reduce the risk of successful social engineering attacks by informing and sensitizing employees about various security policies and security topics, thus ensuring compliance from each individual.
B. Training individuals in security incident response targets is a corrective control action and not as important as proactively preventing an incident.
C. Informing business units about the security strategy is best done through steering committee meetings or other forums.
D. Maintaining evidence of training records to ensure compliance is an administrative, documentary task, but should not be the objective of training.

R1-16 Which of the following is the **GREATEST** risk of a policy that inadequately defines data and system ownership?

A. Audit recommendations may not be implemented.
B. Users may have unauthorized access to originate, modify or delete data.
C. User management coordination does not exist.
D. Specific user accountability cannot be established.

B is the correct answer.

Justification:
A. A policy that inadequately defines data and system ownership generally does not affect the implementation of audit recommendations, particularly because audit reports assign remediation owners.
B. Without a policy defining who has the responsibility for granting access to specific data or systems, there is an increased risk that one could gain (be given) system access without a justified business need. There is a better chance that business objectives will be properly supported when authority to grant access is assigned to specific individuals.
C. While a policy that inadequately defines data and system ownership may affect user management coordination, the greatest risk would be inappropriate granting of user access.
D. User accountability is established by assigning unique user IDs and tracking transactions.

R1-17 A lack of adequate controls represents:

A. a vulnerability.
B. an impact.
C. an asset.
D. a threat.

A is the correct answer.

Justification:
A. The lack of adequate controls represents a vulnerability, exposing sensitive information and data to the risk of malicious damage, attack or unauthorized access by hackers. This could result in a loss of sensitive information, financial loss, legal penalties, etc.
B. Impact is the measure of the financial loss that a threat event may have.
C. An asset is something of either tangible or intangible value worth protecting, including people, systems, infrastructure, finances and reputation.
D. A threat is a potential cause of an unwanted incident.

R1-18 The **PRIMARY** focus of managing IT-related business risk is to protect:

A. information.
B. hardware.
C. applications.
D. databases.

A is the correct answer.

Justification:
A. The primary objective for any enterprise is to protect their mission-critical information based on a risk assessment.
B. While many enterprises spend large amounts protecting their IT hardware, doing so without assessing the risk to their mission-critical data is not advisable. Hardware may become a focus if it stores, processes or transfers mission-critical data.
C. Applications become a focus only if they process mission-critical data.
D. Databases become a focus only if they store mission-critical data.

R1-19 Which of the following provides the **BEST** view of risk management?

A. An interdisciplinary team
B. A third-party risk assessment service provider
C. The enterprise's IT department
D. The enterprise's internal compliance department

A is the correct answer.

Justification:

A. Having an interdisciplinary team contribute to risk management ensures that all areas are adequately considered and included in the risk assessment processes to support an enterprisewide view of risk.
B. Engaging a third party to perform a risk assessment may provide additional expertise to conduct the risk assessment; but without internal knowledge, it will be difficult to assess the adequacy of the risk assessment performed.
C. A risk assessment performed by the enterprise's IT department is unlikely to reflect the view of the entire enterprise.
D. The internal compliance department ensures the implementation of risk responses based on the requirement of management. It generally does not take an active part in implementing risk responses for items that do not have regulatory implications.

R1-20 Which of the following approaches to corporate policy **BEST** supports an enterprise's expansion to other regions, where different local laws apply?

A. A global policy that does not contain content that might be disputed at a local level
B. A global policy that is locally amended to comply with local laws
C. A global policy that complies with law at corporate headquarters and that all employees must follow
D. Local policies to accommodate laws within each region

B is the correct answer.

Justification:

A. Having one global policy that attempts to address local requirements for all locales is nearly impossible and generally cost prohibitive.
B. This choice is correct because it is the only way to minimize the effort and also be in line with local laws.
C. Developing a policy that exclusively takes into account the laws of the corporate headquarters and that does not take into account local laws and regulations will expose the enterprise to various legal actions as well as cause political and reputational loss issues.
D. Local policies for each region, otherwise known as a decentralized approach, requires the enterprise to separately maintain and test a set of documentation/processes for each region, which is extremely expensive. It also does not give an enterprise the opportunity to leverage a set of common practices as is possible by having a global policy that is amended locally.

R1-21 Which of the following is the **BEST** indicator that incident response training is effective?

A. Decreased reporting of security incidents to the incident response team
B. Increased reporting of security incidents to the incident response team
C. Decreased number of password resets
D. Increased number of identified system vulnerabilities

B is the correct answer.

Justification:
A. Decreased reporting is a sign that users are unaware of what constitutes a security incident.
B. Increased reporting of incidents is a good indicator of user awareness, but increased reporting of valid incidents is the best indicator because it is a sign that users are aware of the security rules and know how to report incidents. It is the responsibility of the IT function to assess the information provided, identify false-positives, educate end users, and respond to potential problems.
C. A decrease in the number of password resets is not an indicator of security awareness training.
D. An increase in the number of system vulnerabilities is not an indicator of security awareness training.

R1-22 Which of the following factors will have the **GREATEST** impact on the type of information security governance model that an enterprise adopts?

A. The number of employees
B. The enterprise's budget
C. The organizational structure
D. The type of technology that the enterprise uses

C is the correct answer.

Justification:
A. The number of employees has less impact on information security governance models because well-defined processes provide the proper governance.
B. Organizational budget does not have a major impact on an enterprise's choice of an information security governance model.
C. Information security governance models are highly dependent on the overall organizational structure.
D. The type of technology that the enterprise uses has less impact on information security governance models because well-defined processes provide the proper governance.

R1-23 An enterprise has learned of a security breach at another entity that utilizes similar technology. The **MOST** important action a risk practitioner should take is to:

A. assess the likelihood of the incident occurring at the risk practitioner's enterprise.
B. discontinue the use of the vulnerable technology.
C. report to senior management that the enterprise is not affected.
D. remind staff that no similar security breaches have taken place.

A is the correct answer.

Justification:
A. The risk practitioner should first assess the likelihood of a similar incident occurring at his/her enterprise, based on available information.
B. Discontinuing the use of the vulnerable technology is not necessarily required; furthermore, the technology is likely to be needed to support the enterprise.
C. Reporting to senior management that the enterprise is not affected is premature until the risk practitioner can first assess the impact of the incident.
D. Until research has been conducted, it is not certain that no similar security breaches have taken place.

R1-24 Which of the following is the **GREATEST** benefit of a risk-aware culture?

A. Issues are escalated when suspicious activity is noticed.
B. Controls are double-checked to anticipate any issues.
C. Individuals communicate with peers for knowledge sharing.
D. Employees are self-motivated to learn about costs and benefits.

A is the correct answer.

Justification:

A. Management will benefit most from an escalation process because they can become aware of risk or incidents in a timely manner. In addition, escalation posture among employees will best be trained through a program such as awareness.

B. Double-checking controls is a thorough business practice. It is a basic business stance so the benefit on the management side may be limited.

C. Knowledge sharing is an important theme and this idea should be disseminated through an awareness program. However, the benefit on the risk management side may be indirect.

D. Giving employees a mind-set to learn is desirable. However, cost and benefit knowledge enrichment may not be the primary objective that management expects from awareness efforts.

R1-25 The **MAIN** objective of IT risk management is to:

A. prevent loss of IT assets.
B. provide timely management reports.
C. ensure regulatory compliance.
D. enable risk-aware business decisions.

D is the correct answer.

Justification:

A. Protection of IT assets is a subset goal targeted in an IT risk management program.

B. It is true that an adequate IT risk management program adds value to management reports; an example is presentation of a measurable return of IT investment. However, timeliness in reporting is a separate issue to be discussed apart from IT risk management.

C. Meeting regulatory compliance requirements is a part of the objectives to be achieved in an IT risk management framework.

D. IT risk management should be conducted as part of enterprise risk management (ERM), the ultimate objective of which is to enable risk-aware business decisions.

R1-26 Which of the following is the **BEST** risk identification technique for an enterprise that allows employees to identify risk anonymously?

A. The Delphi technique
B. Isolated pilot groups
C. A strengths, weaknesses, opportunities and threats (SWOT) analysis
D. A root cause analysis

A is the correct answer.

Justification:

A. With the Delphi technique, polling or information gathering is done either anonymously or privately between the interviewer and interviewee.
B. With isolated pilot groups, participants generally do not anonymously identify risk.
C. With a strengths, weaknesses, opportunities and threats (SWOT) analysis, participants generally do not anonymously identify risk.
D. With a root cause analysis, participants generally do not anonymously identify risk.

R1-27 Who **MUST** give the final sign-off on the IT risk management plan?

A. IT auditors performing the risk assessment
B. Business process owners
C. Senior management
D. IT security administrators

C is the correct answer.

Justification:

A. IT auditors performing the risk assessment may be involved in creating the risk management plan, but they do not have the authority to give the final sign-off.
B. Business process owners may be involved in creating the risk management plan, but they do not have the authority to give the final sign-off.
C. By understanding the system and subsystem, senior management have knowledge of the performance metrics and indicators used to measure the system and subsystems, of how the policies and standards are applied within the system and subsystems, and an understanding of the risk and potential impacts associated with recent audit findings and recommendations.
D. IT security administrators may be involved in creating the risk management plan, but they do not have the authority to give the final sign-off.

R1-28 Which of the following is the **PRIMARY** reason that a risk practitioner determines the security boundary prior to conducting a risk assessment?

A. To determine which laws and regulations apply
B. To determine the scope of the risk assessment
C. To determine the business owner(s) of the system
D. To decide between conducting a quantitative or qualitative analysis

B is the correct answer.

Justification:
A. The risk assessment itself will take into account the laws and regulations that apply.
B. The primary reason for determining the security boundary is to establish what systems and components are included in the risk assessment.
C. Determining the business owners is a reason for determining the security boundary, but is secondary to determining the scope.
D. The security boundary should not be a primary factor in deciding between conducting a quantitative or qualitative risk analysis.

R1-29 Which of the following **BEST** describes the information needed for each risk on a risk register?

A. Various risk scenarios with their date, description, impact, probability, risk score, mitigation action and owner
B. Various risk scenarios with their date, description, risk score, cost to remediate, communication plan and owner
C. Various risk scenarios with their date, description, impact, cost to remediate and owner
D. Various activities leading to risk management planning

A is the correct answer.

Justification:
A. This choice is the best answer because it contains the necessary elements of the risk register that are needed to make informed decisions.
B. This choice contains some elements of a risk register, but misses some important and key elements of a risk register (impact, probability, mitigation action) that are needed to make informed decisions and this choice lists some items that should not be included in the register (communication plan).
C. This choice misses some important and key elements of a risk register (probability, risk score, mitigation action) needed to make informed decisions.
D. A risk register is a result of risk management planning, not the other way around.

R1-30 The **GREATEST** advantage of performing a business impact analysis (BIA) is that it:

A. does not have to be updated because the impact will not change.
B. promotes continuity awareness in the enterprise.
C. can be performed using only qualitative estimates.
D. eliminates the need to perform a risk analysis.

B is the correct answer.

Justification:
A. A business impact analysis (BIA) will need to be updated periodically because systems do change and new systems are added.
B. A BIA raises the level of awareness for business continuity within the enterprise.
C. A BIA should utilize both qualitative and quantitative estimates; however, the analysis can still be completed and estimates determined with or without minimum historical data.
D. A BIA does not eliminate the need to perform a risk analysis, and although it is a part of the documentation used during a risk analysis, it is the not the greatest advantage.

R1-31 The **PRIMARY** advantage of creating and maintaining a risk register is to:

A. ensure that an inventory of potential risk is maintained.
B. record all risk scenarios considered during the risk identification process.
C. collect similar data on all risk identified within the organization.
D. run reports based on various risk scenarios.

A is the correct answer.

Justification:
A. Once important assets and the risk that may impact these assets are identified, the risk register is used as an inventory of that risk. The risk register can help enterprises accelerate their risk decision making and establish accountability for specific risk.
B. It is good practice to record all considered scenarios in the register and reassess them annually; however, maintaining the inventory is the primary advantage.
C. A spreadsheet or governance, risk and compliance (GRC) tool allows for the collection of similar data elements in a single format, but ensuring the inventory is still the primary advantage.
D. The ability to run reports is a benefit of having a risk registry, but not its primary purpose.

R1-32 Which of the following is **MOST** effective in assessing business risk?

A. A use case analysis
B. A business case analysis
C. Risk scenarios
D. A risk plan

C is the correct answer.

Justification:
A. A use case analysis is the most common technique used to identify the requirements of a system and the information used to define processes.
B. Business cases are generally a part of the project charter and help define the purpose/reason for the project.
C. Risk scenarios are the most effective technique in assessing business risk.
D. A risk plan is the output document from the assessment of risk.

R1-33 The board of directors of a one-year-old start-up company has asked their chief information officer (CIO) to create all of the enterprise's IT policies and procedures. Which of the following should the CIO create **FIRST**?

A. The strategic IT plan
B. The data classification scheme
C. The information architecture document
D. The technology infrastructure plan

A is the correct answer.

Justification:

A. The strategic IT plan is the first policy to be created when setting up an enterprise's governance model.

B. The strategic IT plan is created before the data classification scheme is developed. The data classification scheme is a method for classifying data by factors such as criticality, sensitivity and ownership.

C. The strategic IT plan is created before the information architecture is defined. The information architecture is one component of the IT architecture (together with applications and technology). The IT architecture is a description of the fundamental underlying design of the IT components of the business, the relationships among them, and the manner in which they support the organization's objectives.

D. The strategic IT plan is created before the technology infrastructure plan is developed. The technology infrastructure plan maps out the technology, human resources and facilities that enable the current and future processing and use of applications.

R1-34 The preparation of a risk register begins in which risk management process?

A. Risk response planning
B. Risk monitoring and control
C. Risk management planning
D. Risk identification

D is the correct answer.

Justification:

A. In the risk response planning process, appropriate responses are chosen, agreed-on and included in the risk register.

B. Risk monitoring and control often requires identification of new risk and reassessment of risk. Outcomes of risk reassessments, risk audits and periodic risk reviews trigger updates to the risk register.

C. Risk management planning describes how risk management will be structured and performed.

D. The risk register details all identified risk, including description, category, cause, probability of occurring, impact(s) on objectives, proposed responses, owners and current status. The primary outputs from risk identification are the initial entries into the risk register.

R1-35 A business impact analysis (BIA) is **PRIMARILY** used to:

A. estimate the resources required to resume and return to normal operations after a disruption.
B. evaluate the impact of a disruption to an enterprise's ability to operate over time.
C. calculate the likelihood and impact of known threats on specific functions.
D. evaluate high-level business requirements.

B is the correct answer.

Justification:

A. Determining the resource requirements to resume and return to normal operations is part of business continuity planning.
B. A business impact analysis (BIA) is primarily used for evaluating the impact of a disruption over time to an enterprise's ability to operate. It determines the urgency of each business activity. Key deliverables include recovery time objectives (RTOs) and recovery point objectives (RPOs).
C. Likelihood and impact are calculated during risk analysis.
D. High-level business requirements are defined during the early phases of a system development life cycle (SDLC), not as part of a BIA.

R1-36 Which of the following provides the **GREATEST** level of information security awareness?

A. Job descriptions
B. A security manual
C. Security training
D. An organizational diagram

C is the correct answer.

Justification:

A. While job descriptions are useful to describe job-related roles and responsibilities, including those related to security, they do not provide sufficient detail to enable employees who are not already proficient in security to understand how they can actively support and contribute to a risk-aware culture. Such detail is generally only provided through security training.
B. A security manual is written for a technical audience and is usually not accessible to all staff. For this reason it is not a viable mechanism to make employees aware of their security responsibilities.
C. Security training is the best way to inform all employees about information security awareness.
D. An organizational diagram shows the various departmental hierarchies, but it is not associated to information security awareness. It can, however, be used by the information security team to determine which individuals need what type of information security awareness.

R1-37 Which of the following is the **BIGGEST** concern for a chief information security officer (CISO) regarding interconnections with systems outside of the enterprise?

A. Requirements to comply with each other's contractual security requirements
B. Uncertainty that the other system will be available as needed
C. The ability to perform risk assessments on the other system
D. Ensuring that communication between the two systems is encrypted through a virtual private network (VPN) tunnel

A is the correct answer.

Justification:

A. Ensuring that both systems comply with the contractual security requirements of both entities should be the primary concern of the risk practitioner. If one system falls out of compliance, then they both most likely will miss their respective security requirements.
B. Uncertainty of the other system's availability is probably the primary concern of the business owner and users, not of the chief information security officer (CISO).
C. The ability to perform risk assessment on the other system may or may not be a concern based on the interconnection agreement between the two systems.
D. Communication between the two systems may not necessarily require a virtual private network (VPN) tunnel, or encryption. That requirement will be based on type of data being transmitted.

R1-38 The board of directors of a one-year-old start-up company has asked their chief information officer (CIO) to create all of the enterprise's IT policies and procedures, which will be managed and approved by the IT steering committee. The IT steering committee will make all of the IT decisions for the enterprise, including those related to the technology budget. Which type of IT organizational structure does the enterprise have?

A. Project-based
B. Centralized
C. Decentralized
D. Divisional

B is the correct answer.

Justification:

A. This choice is incorrect because a project-based enterprise is one where a group is formed temporarily to work on one particular project. The group that the chief information officer (CIO) is setting up, and steering committees in general, is not temporary.
B. With a centralized IT organizational structure, all of the decisions are made by one group for the entire enterprise.
C. This choice is incorrect because with a decentralized organizational structure decisions are made by each division (sales, human resources [HR], etc.). of the organization. If this were to occur, then there would be several different, and perhaps conflicting, IT policies.
D. This choice is incorrect because with a divisional organizational structure each geographic area or product or service will have its own group.

R1-39 Which of the following is the **MAIN** concern when two or more staff members are allowed to use the same generic account?

A. Segregation of duties
B. Inability to change the password
C. Repudiation
D. Inability to trace account activities

C is the correct answer.

Justification:

A. It is not an issue of segregation of duties because the user may be performing identical activities. Because they have the same login credentials, it is a repudiation issue: none of the users can be held accountable because each user can deny accountability for transactions performed under the generic account.

B. System password parameters or rules are independent of users sharing generic accounts. Specific parameters such as password history will be effective.

C. Because the username and password are the same in the case of a generic account, this will impact the nonrepudiation of information; it will be difficult to establish which user logged in and performed the various operations. However, with the right tools the activity can be traced back to the media access control (MAC) address if users are accessing information through different terminals. Repudiation is the denial of a transaction by one of the parties, or of participation in all or part of that transaction, or of the content of communication related to that transaction.

D. Activities can be traced by generic user name by enabling the system auditing or similar functionalities.

R1-40 Which of the following is a **PRIMARY** consideration when developing an IT risk awareness program?

A. Why technology risk is owned by IT
B. How technology risk can impact each attendee's area of business
C. How business process owners can transfer technology risk
D. Why technology risk is more difficult to manage compared to other risk

B is the correct answer.

Justification:

A. IT does not own technology risk. An appropriate topic of IT risk awareness training may be the fact that many types of IT risk are owned by the business. One example may be the risk of employees exploiting insufficient segregation of duties within an enterprise resource planning (ERP) system.

B. Stakeholders must understand how the IT-related risk impacts the overall business.

C. Transferring risk is not of primary consideration in developing a risk awareness program. It is a part of the risk response process.

D. Technology risk may or may not be more difficult to manage than other types of risk. Although this is important from an awareness point of view, it is not as primary as understanding the impact in the area of business.

R1-41 Which of the following is the **BEST** approach when conducting an IT risk awareness campaign?

A. Provide technical details on exploits.
B. Provide common messages tailored for different groups.
C. Target system administrators and help desk staff.
D. Target senior managers and business process owners.

B is the correct answer.

Justification:
A. Providing technical details on exploits is not advisable during an IT risk awareness campaign because this could teach individuals how to circumvent controls.
B. Groups have differing levels of responsibility and expertise; each group should receive the same message tailored to its role and level of understanding.
C. Specific groups should not be singled out for training at the exclusion of others because all groups have a role to play in strengthening information systems security.
D. Specific groups should not be singled out for training at the exclusion of others because all groups have a role to play in strengthening information systems security.

R1-42 It is **MOST** important that risk appetite be aligned with business objectives to ensure that:

A. resources are directed toward areas of low risk tolerance.
B. major risk is identified and eliminated.
C. IT and business goals are aligned.
D. the risk strategy is adequately communicated.

A is the correct answer.

Justification:
A. Risk appetite is the amount of risk that an enterprise is willing to take on in pursuit of value. Aligning it with business objectives allows an enterprise to evaluate and deploy valuable resources toward those objectives where the risk tolerance (for loss) is low.
B. There is no link between aligning risk appetite with business objectives and identification and elimination of major risk. Moreover, risk cannot be eliminated; it can be reduced to an acceptable level using various risk response options.
C. Alignment of risk appetite with business objectives does converge IT and business goals to a point, but alignment is not limited to these two areas. Other areas include organizational, strategic and financial objectives, among other objectives.
D. Communication of the risk strategy does not depend on aligning risk appetite with business objectives.

R1-43 Risk scenarios enable the risk assessment process because they:

A. cover a wide range of potential risk.
B. minimize the need for quantitative risk analysis techniques.
C. segregate IT risk from business risk for easier risk analysis.
D. help estimate the frequency and impact of risk.

D is the correct answer.

Justification:
A. The use of risk scenarios does not indicate that the enterprise is assessing a wide range of risks.
B. Risk scenarios do not necessarily minimize the need for quantitative risk analysis.
C. Risk scenarios can be applied to both IT risk and business risk and there is no question of segregating the risk.
D. Because risk scenarios cover a wide/broad range of potential risk, assessment is simplified using various threats and vulnerabilities for mapping in a practical context.

R1-44 Who is accountable for business risk related to IT?

A. The chief information officer (CIO)
B. The chief financial officer (CFO)
C. Users of IT services—the business
D. The chief architect

C is the correct answer.

Justification:
A. The chief information officer (CIO) supports the business, but does not own the business risk.
B. The chief financial officer (CFO) tracks the costs of the resources and the financial risk, but does not own the business risk.
C. Ultimately, it is the business—the users of IT services—who must own business-related risk, including the risk related to the use of IT. The business should set the mandate for risk management, provide the resources and funding to support a risk management plan designed to protect business interests, and monitor whether risk is being managed.
D. The chief architect mitigates IT risk through the architecture of the IT environment, but does not own the business risk.

R1-45 Which of the following information in the risk register **BEST** helps in developing proper risk scenarios? A list of:

A. potential threats to assets.
B. residual risk on individual assets.
C. accepted risk.
D. security incidents.

A is the correct answer.

Justification:
A. Potential threats that may impact the various business assets will help in developing scenarios on how these threats can exploit vulnerabilities and cause a risk and therefore help in developing proper risk scenarios.
B. Residual risk on individual assets does not help in developing a proper risk scenario.
C. Accepted risk is generally a small subset of entries within the risk register. Accepted risk should be included in the risk register to ensure that events that may affect the current decision of the enterprise to accept the risk are monitored.
D. Previous security incidents of the enterprise itself or entities with a similar profile may inspire similar risk scenarios to be included in the risk register. However, the best approach to create a meaningful risk register is to capture potential threats on tangible and intangible assets.

R1-46 Which of the following is true about IT risk?

A. IT risk cannot be assessed and measured quantitatively.
B. IT risk should be calculated separately from business risk.
C. IT risk management is the responsibility of the IT department.
D. IT risk exists whether or not it is detected or recognized by an enterprise.

D is the correct answer.

Justification:
A. IT risk, like any business risk, can be assessed both quantitatively and qualitatively. It is very difficult and incomplete to measure risk quantitatively.
B. IT risk is one type of business risk.
C. IT risk is the responsibility of senior management, not just the IT department.
D. The enterprise must identify, acknowledge and respond to risk; ignorance of risk is not acceptable.

R1-47 Which of the following is **MOST** important when selecting an appropriate risk management methodology?

A. Risk culture
B. Countermeasure analysis
C. Cost-benefit analysis
D. Risk transfer strategy

A is the correct answer.

Justification:
A. Without an understanding of the risk culture, in terms of how the enterprise makes decisions regarding risk, one cannot select a risk management methodology.
B. Countermeasure analysis is used for analyzing those controls that address specific attacks, at times while the attack is occurring. Countermeasure analysis is not a factor when selecting an appropriate risk management methodology.
C. Cost-benefit analysis is used to measure the projected benefit of a solution, such as a control, against its price. The analysis can be at a point in time or over an extended period. It is generally not a consideration when selecting a risk management methodology.
D. Because not all risk can be transferred, implementing a proper risk assessment methodology must begin with considering the overall risk profile not with the risk transfer strategy.

R1-48 Which of the following **BEST** determines compliance with the risk appetite of an enterprise?

A. Balance between preventive and detective controls
B. Inherent risk and acceptable risk level
C. Residual risk and acceptable risk level
D. Balance between countermeasures and preventive controls

C is the correct answer.

Justification:

A. Balance between preventive and detective controls does not help evaluate the risk appetite because these controls may have been established as a result of risk analysis at that point in time.
B. Inherent risk in itself does not help in knowing the residual risk, and that in combination with acceptable risk level is inadequate information to evaluate the risk appetite of an enterprise.
C. Residual risk along with acceptable risk level will provide information on risk that is balanced after the application of controls. Management can decide whether to accept the risk or apply more controls based on acceptable risk levels. This will also help to understand the risk appetite of the enterprise because a conservative approach tries to accept risk levels that are low or very low. So this is the primary information that will be useful to management for making risk-related decisions.
D. Countermeasures are put in place when the threat needs to be further reduced and do not help in evaluating the risk appetite of an enterprise.

R1-49 Which of the following **BEST** improves decision making related to risk?

A. Maintaining a documented risk register of all possible risk
B. Risk awareness training in line with the risk culture
C. Maintaining updated security policies and procedures
D. Allocating accountability of risk to the department as a whole

A is the correct answer.

Justification:

A. Maintaining a documented risk register improves decision making related to risk response because a risk register captures the population of relevant risk scenarios and helps provide a basis for prioritization of risk responses.
B. Imparting risk awareness training to various stakeholders and customizing the contents according to the enterprise's risk culture will sensitize the stakeholders and users responsible for risk responsibilities and accountabilities to make decisions on acceptance of residual risk.
C. Maintaining policies and procedures will not improve the decision making related to residual risk.
D. Allocating accountability to the department as a whole will circumvent ownership because there will be no individual owner for risk.

R1-50 The **FIRST** step in identifying and assessing IT risk is to:

A. confirm the risk tolerance level of the enterprise.
B. identify threats and vulnerabilities.
C. gather information on the current and future environment.
D. review past incident reports and response activity.

C is the correct answer.

Justification:
A. A risk practitioner must understand the risk appetite of senior management and the associated risk tolerance level. This is not the first step because risk tolerance becomes relevant during risk response.
B. Identification of relevant threats and vulnerabilities is important, but is limited in its view.
C. The first step in any risk assessment is to gather information about the current state and pending internal and external changes to the enterprise's environment (scope, technology, incidents, modifications, etc.).
D. While the review of past incident reports may be an input for the identification and assessment of IT risk, focusing on these factors is not prudent.

R1-51 Which of the following outcomes of an outsourcing contract for non-core processes is of **GREATEST** concern to the management of an enterprise?

A. Total cost of ownership (TCO) exceeds projections.
B. Internal information systems experience has been lost.
C. Employees of the vendor were disloyal to the client enterprise.
D. Processing of critical data was subcontracted by the vendor.

D is the correct answer.

Justification:
A. Total cost of ownership (TCO) exceeding projections is significant, but not uncommon. Because TCO is based on modeling, some variation can be expected.
B. The loss of internal information systems experience is often a concern when processes or subprocesses that are considered "core" are outsourced. However, for non-core processes, the loss of such experience would not be a concern.
C. Lack of vendor loyalty to the client enterprise is generally managed via service level agreements (SLAs).
D. The greatest risk in third-party relationships is the fact that the enterprise is losing control of its IS processes to the vendor. Subcontracting will increase this risk; therefore, the subcontracting process has to be reviewed because critical data are involved.

R1-52 Which of the following is **MOST** important for effective risk management?

A. Assignment of risk owners to identified risk
B. Ensuring compliance with regulatory requirements
C. Integration of risk management into operational processes
D. Implementation of a risk avoidance strategy

A is the correct answer.

Justification:
A. It is of utmost importance to assign identified risk to owners who will be responsible for the risk.
B. Regulatory compliance is a small part of the risk management process.
C. Risk management should be integrated into strategic, tactical and operational processes of an enterprise.
D. Risk avoidance is not always feasible in a business environment.

R1-53 Risk scenarios should be created **PRIMARILY** based on which of the following?

A. Input from senior management
B. Previous security incidents
C. Threats that the enterprise faces
D. Results of the risk analysis

C is the correct answer.

Justification:
A. Input from senior management is not as critical as organizational threats.
B. Previous incidents are not as critical as organizational threats.
C. When creating risk scenarios the most important factor to consider is the likely threats or threat actions that will act upon the risk.
D. Results of the risk analysis should not be available until after risk scenarios are created because the scenarios should be an input into the risk analysis.

R1-54 Which of the following causes an internal *ad hoc* risk assessment to be performed before the annual occurrence?

A. A new chief information officer (CIO) is hired.
B. Senior management adjusts risk appetite.
C. Risk changes on a frequent basis.
D. A new system is introduced into the environment.

D is the correct answer.

Justification:
A. A new chief information officer (CIO) may require a new organizational risk assessment, but it would not be required because he/she could review the last risk assessment if there have been no changes to the environment.
B. Senior management adjusting risk appetite will significantly affect risk responses, but does not require a risk assessment.
C. Risk changing on a frequent basis will be captured during the annual risk assessment.
D. Once a new system is added into the enterprise it adds to the overall risk to that enterprise's business objectives. The amount of risk should be determined via a risk assessment.

R1-55 An enterprise expanded operations into Europe, Asia and Latin America. The enterprise has a single-version, multiple-language employee handbook last updated three years ago. Which of the following is of **MOST** concern?

A. The handbook may not have been correctly translated into all languages.
B. Newer policies may not be included in the handbook.
C. Expired policies may be included in the handbook.
D. The handbook may violate local laws and regulations.

D is the correct answer.

Justification:

A. Translation errors may lead to confusion and lack of policy enforceability in a given region, but it still is generally considered less damaging than including policies that defy local laws and customs.
B. While the handbook may not reflect the latest policies, it is more important to be in compliance with local laws and regulations.
C. While the handbook may contain expired policies, it is more important to be in compliance with local laws and regulations.
D. Because customs and laws play a role in an enterprise's ability to effectively operate in a given location (both varying by state and by country), it is important for the employee handbook to appropriately acknowledge and account for such differences.

R1-56 When requesting information for an e-discovery, an enterprise learned that their email cloud provider was never contracted to back up the messages even though the company's email retention policy explicitly states that all emails are to be saved for three years. Which of the following would have **BEST** safeguarded the company from this outcome?

A. Providing the contractor with the record retention policy up front
B. Validating the company policies to the provider's contract
C. Providing the contractor with the email retention policy up front
D. Backing up the data on the company's internal network nightly

B is the correct answer.

Justification:

A. Providing the contractor with the record retention policy still does not legally bind the third party to perform the activities in accordance with internal policies.
B. The initial review of any third-party service should include a validation that the vendor is contractually required to abide by internal policies, including record retention if the enterprise's record retention policy specifically calls for data that will be managed by a third party.
C. Providing the contractor with the email record retention policy in and of itself does not legally bind the third party to perform the activities in accordance with internal policies.
D. An enterprise can choose to perform internal backups on the data stored by a third party. Although this may give the correct retention period, during e-discovery the opposing party may ask for the original source and at that point the policy variance would become apparent.

R1-57 Which of the following is the **BEST** indicator of an effective information risk management program?

A. The security policy is made widely available.
B. Risk is considered before all decisions.
C. Security procedures are updated annually.
D. Risk assessments occur on an annual basis.

B is the correct answer.

Justification:
A. Making the security policy widely available will assist in ensuring the success, but is not as critical as making risk-based business decisions.
B. Ensuring that risk is considered and determined before business decisions are made best ensures that risk tolerance is kept at the level approved by the organization.
C. Updating security procedures annually is only necessary if policy changes.
D. Ensuring that risk assessments occur annually will assist in ensuring success, but is not as critical as making risk-based business decisions.

R1-58 Risk management strategic plans are **MOST** effective when developed for:

A. the enterprise as a whole.
B. each individual system based on technology utilized.
C. every location based on geographic threats.
D. end-to-end business processes.

A is the correct answer.

Justification:
A. Risk management strategic plans are most effective when they are created and followed by the entire enterprise.
B. Because most enterprises use a great variety of technologies, creating a management plan for each technology utilized in the enterprise leads to having too many plans to follow, and some of the policies may conflict with others.
C. It is difficult to create a risk management plan for each location based on geographic threats. Also, these plans do not take other types of threats into account.
D. Having risk management plans based on end-to-end business processes leads to confusion and overlapping and conflicting policies and procedures.

R1-59 Which of the following is **MOST** important when considering the risk appetite of an enterprise?

A. The capacity of the enterprise to absorb loss
B. The definition of responsibilities for risk management
C. The line of business and the typical risk of the industry
D. The culture and predisposition toward risk taking

D is the correct answer.

Justification:

A. While the capacity of the enterprise to absorb loss is an important risk mitigation factor, it does not influence risk appetite as much as culture and predisposition toward risk taking.
B. The definition of the responsibilities and the accountability for IT risk management says nothing about the enterprise's risk appetite. Risk appetite mostly depends on the risk culture, the risk tolerance and risk acceptance of an enterprise.
C. The line of business and typical risk of the industry say nothing about the risk appetite of a single enterprise. Risk appetite is an individual aspect and depends on the individual risk culture of an enterprise.
D. When considering risk appetite, two major factors are relevant: the management culture and the predisposition toward risk taking.

R1-60 The **PRIMARY** purpose of adopting an enterprisewide risk management framework is to:

A. allow the flexibility to adjust the risk response strategy throughout the enterprise.
B. centralize the responsibility for the maintenance of the risk response program.
C. enable a consistent approach to risk response throughout the enterprise.
D. avoid higher costs for risk reduction and audit strategies throughout the enterprise.

C is the correct answer.

Justification:

A. A risk management framework enables a consistent approach while allowing the necessary flexibility at the local level.
B. Risk management is the responsibility of all individuals. Accountability of risk management lies with senior management and the board.
C. Enabling a consistent approach to risk response is a key objective of a risk management framework.
D. Avoiding higher costs for risk reduction and audit strategies throughout the enterprise is good practice, but not the primary purpose of adopting an enterprisewide risk management framework.

R1-61 A review of an enterprise's IT projects finds that projects frequently go over time or budget by nearly 10 percent. On review, management advises the risk practitioner that a deviation of 15 percent is acceptable. This is an example of:

A. risk avoidance.
B. risk tolerance.
C. risk acceptance.
D. risk mitigation.

B is the correct answer.

Justification:

A. Risk avoidance involves terminating or suspending an activity to steer clear of the inherent risk. Risk avoidance generally also affects the potential opportunity offered by engaging in the activity.
B. Risk tolerance is the permissible deviation from declared risk appetite levels.
C. Risk acceptance means that the enterprise makes an educated decision not to take action relative to a particular risk and accepts loss when/if it occurs.
D. Risk mitigation is the management of risk through the use of countermeasures and controls.

R1-62 When a start-up company becomes popular, it suddenly is the target of hackers. This is considered:

A. an emerging vulnerability.
B. a vulnerability event.
C. an emerging threat.
D. an environmental risk factor.

C is the correct answer.

Justification:

A. A vulnerability is a weakness in the design, implementation, operation or internal control of a process that can expose the system to adverse threats from threat events, which is not described in the question stem.
B. A vulnerability event is any event from which a material increase in vulnerability results from changes in control conditions or from changes in threat capability/force.
C. A threat is any event in which a threat element/actor acts against an asset in a manner that has the potential to directly result in harm. The stem describes the emerging threat of hackers attacking the start-up company.
D. Environmental risk factors can be split into internal and external environmental risk factors. Internal environmental factors are, to a large extent, under the control of the enterprise, although they may not always be easy to change. External environmental factors are, to a large extent, outside the control of the enterprise. The question stem best describes the emerging threat of hackers attacking the start-up company.

R1-63 Which of the following is a **MAJOR** risk associated with the use of governance, risk and compliance (GRC) tools?

A. Misinterpretation of the dashboard's output
B. Poor authentication mechanism
C. Obsolescence of content
D. Complex integration of the diverse requirements

C is the correct answer.

Justification:
A. Misinterpreting the dashboard's output can be overcome easily by training or using subject matter experts.
B. New technologies have overcome the challenge of poor authentication mechanisms.
C. A governance, risk and compliance (GRC) application has to be updated regularly with current regulations, policies, etc. Obsolete content will render the GRC outdated. Many GRC applications are based on the unified compliance framework (UCF) for mapping to various regulations, frameworks and standards. The technology team needs to refresh the UCF file quarterly through its vendor and needs to have a process to identify and address all changes from one release to the next. Additionally, the organization needs to commit internal resources to maintain all the company data in the tool to keep the content from obsolescence.
D. Integrating the diverse, complex requirements is a challenge that most GRC tools have been designed to overcome. Choice C is more serious because it is entirely in the hands of the users.

R1-64 The **PRIMARY** reason an external risk assessment team reviews documentation before starting the actual risk assessment is to gain a thorough understanding of:

A. the technologies utilized.
B. gaps in the documentation.
C. the enterprise's business processes.
D. the risk assessment plan.

C is the correct answer.

Justification:
A. Reviewing the technology used can be conducted during the risk assessment.
B. Identifying gaps in the documentation can occur during the risk assessment.
C. The risk assessment team should thoroughly understand the enterprise's business processes and objectives before the assessment in order to appropriately evaluate risk.
D. The risk assessment plan should be created by the external auditors.

R1-65 A small start-up software development company has been flooded and the insurance does not pay out because the premium has lapsed. In relation to risk management, the lapsed premium is considered a:

A. risk.
B. vulnerability.
C. threat.
D. negligence.

B is the correct answer.

Justification:

A. A risk is the combination of the probability of an event and its consequence (ISO/IEC 73). The question stem describes a weakness in the insurance premium payment process, which is considered a vulnerability.

B. The question stem describes a vulnerability. A vulnerability is a weakness in the design, implementation, operation or internal control of a process that could expose the enterprise to adverse threats from threat events.

C. A threat is anything (e.g., object, substance, human) that is capable of acting against an asset in a manner that can result in harm. In the case of the question stem, the threat is the flood.

D. Negligence is a legal term describing a civil wrong causing injury or harm to another person or to property as the result of doing something or failing to provide a proper or reasonable level of care. Negligence is not specifically related to risk management.

R1-66 Which of the following statements **BEST** describes the value of a risk register?

A. It captures the risk inventory.
B. It drives the risk response plan.
C. It is a risk reporting tool.
D. It lists internal risk and external risk.

B is the correct answer.

Justification:

A. A risk register is used to provide detailed information on each identified risk such as risk owner, details of the scenario and assumptions, affected stakeholders, causes/indicators, information on the detailed scores (i.e., risk ratings) on the risk analysis, and detailed information on the risk response (e.g., action owner and the risk response status, time frame for action, related projects, and risk tolerance level). These components can also be defined as the risk universe.

B. Risk registers serve as the main reference for all risk-related information, supporting risk-related decisions such as risk response activities and their prioritization.

C. Risk register data are utilized to generate management reports, but are not in themselves a risk reporting tool.

D. The risk register tracks all internal and external risk, the quality and quantity of the controls, and the likelihood and impact of the risk.

R1-67 Accountability for risk ultimately belongs to the:

A. chief risk officer (CRO).
B. compliance officer.
C. chief financial officer (CFO).
D. board of directors.

D is the correct answer.

Justification:
A. The chief risk officer (CRO) has responsibilities in risk management, but is not ultimately accountable for risk.
B. The compliance officer has responsibility for certain risk governance and risk response activities, but is not ultimately accountable.
C. The chief financial officer (CFO) has major responsibilities in risk management, but is not ultimately accountable.
D. The board of directors of an enterprise has ultimate accountability to shareholders, customers, employees and the general public.

R1-68 What is the **MAIN** objective of risk identification?

A. To detect possible threats that may affect the business
B. To ensure that risk factors and root causes are managed
C. To enable the review of the key performance indicators (KPIs)
D. To provide qualitative impact values to stakeholders

A is the correct answer.

Justification:
A. Risk identification is the process of determining and documenting the risk that an enterprise faces. The identification of risk is based on the recognition of threats, vulnerabilities, assets and controls in the enterprise's operational environment.
B. Ensuring that risk factors and root causes are addressed is the objective of the risk response process, not risk identification.
C. Enabling the review of the key performance indicators (KPIs) is the objective of the risk monitoring process.
D. Qualitative risk impact values are derived from the risk assessment process.

R1-69 Which of the following examples of risk should be addressed during application design?

A. A lack of skilled resources
B. The risk of migration to a new system
C. Incomplete technical specifications
D. Third-party supplier risk

A is the correct answer.

Justification:
A. A lack of skilled resources implies that the project is beyond the skills of the personnel involved and is associated with the design phase.
B. Migration risk is typically associated with the implementation phase.
C. Technical risk is introduced when the technical requirements may be beyond the scope of the project.
D. Risk that a third-party supplier would not be able to deliver on time or to requirements is associated with the implementation phase.

R1-70 If risk has been identified, but not yet mitigated, the enterprise would:

A. record and mitigate serious risk and disregard low-level risk.
B. obtain management commitment to mitigate all identified risk within a reasonable time frame.
C. document all risk in the risk register and maintain the status of the remediation.
D. conduct an annual risk assessment, but disregard previous assessments to prevent risk bias.

C is the correct answer.

Justification:

A. All levels of risk identified should be documented in the risk register. It is important to be able to identify where low-level risk can be aggregated within the register.
B. Not all identified risk will automatically be mitigated. The enterprise will conduct a cost-benefit analysis before determining the appropriate risk response.
C. All identified risk should be included in the risk register. The register should capture the proposed remediation plan, the risk owner and anticipated date of completion.
D. Annual risk assessments should consider previous risk assessments.

R1-71 When developing IT-related risk scenarios with a top-down approach, it is **MOST** important to identify the:

A. information system environment.
B. business objectives.
C. hypothetical risk scenarios.
D. external risk scenarios.

B is the correct answer.

Justification:

A. Top-down risk scenario development identifies the enterprise's business objectives and builds risk scenarios based on situations that may affect these objectives. The information system environment would be a risk factor.
B. Typically, top-down risk scenario development is performed by identifying business objectives and identifying risk scenarios with the greatest impact on the achievement of business objectives.
C. The identification of generic risk scenarios is usually related to a bottom-up risk identification method.
D. It is important to identify both external and internal risk scenarios.

R1-72 An enterprise has outsourced several business functions to a firm in another country, including IT development, data hosting and support. What is the **MOST** important consideration the risk professional will examine in relation to the outsourcing arrangements?

A. Are policies and procedures in place to handle security exceptions?
B. Is the outsourcing supplier meeting the terms of the service level agreements (SLAs)?
C. Is the security program of the outsourcing provider based on an international standard?
D. Are specific security controls mandated in the outsourcing contract/agreement?

D is the correct answer.

Justification:
A. There should be policies and procedures to handle incidents or exceptional circumstances; however, this is not the most important consideration.
B. Whether the provider meets the service level agreements (SLAs) is of concern to the outsourcing enterprise and the auditors; however, this is not the most important consideration. Stipulating the SLA in the contract is the first requirement.
C. The contract should stipulate the required levels of security and risk management. Basing the security program on a recognized international standard may be an excellent foundation for the security program, but is not the most important consideration.
D. Without addressing security requirements directly in the outsourcing contract, the outsourcing company has no assurance that the provider will be compliant with specific security requirements.

R1-73 The **MAIN** purpose for creating and maintaining a risk register is to:

A. ensure that all assets have low residual risk.
B. define the risk assessment methodology.
C. document all identified risk.
D. study various risk scenarios in the threat landscape.

C is the correct answer.

Justification:
A. Creating and maintaining the risk register does not automatically ensure that all assets have low residual risk. The reduction in risk is based on mitigating controls.
B. Creating and maintaining a risk register is the result of following a risk assessment methodology.
C. A risk register is used to provide detailed information on each identified risk such as risk owner, details of the scenario and assumptions, affected stakeholders, causes/indicators, information on the detailed scores (i.e., risk ratings) on the risk analysis, detailed information on the risk response (e.g., action owner and the risk response status, time frame for action, related projects, and risk tolerance level). These components can also be defined as the risk universe.
D. Creating and maintaining a risk register is not about studying various risk scenarios and redefining the threat landscape. This can be achieved without creating the risk register.

R1-74 Which of the following activities provides the **BEST** basis for establishing risk ownership?

A. Documenting interdependencies between departments
B. Mapping identified risk to a specific business process
C. Referring to available RACI charts
D. Distributing risk equally among all asset owners

B is the correct answer.

Justification:

A. Documenting interdependencies between departments helps identify the work flow, but does not identify risk ownership.
B. Mapping identified risk to a specific business process helps identify the process owner. Aggregation of related business processes results in identification of the prospective risk owner.
C. The review of a RACI chart identifies who is responsible, accountable, consulted and informed within an organizational framework, but does not identify risk ownership.
D. Ownership of risk cannot be a shared responsibility, but each risk must be allocated to specific owners.

R1-75 Which of the following types of risk is high for projects that affect multiple business areas?

A. Control risk
B. Inherent risk
C. Compliance risk
D. Residual risk

B is the correct answer.

Justification:

A. Control risk could be high, but it would be due to internal controls not being identified, evaluated or tested and would not be due to the number of users or business areas affected.
B. Inherent risk is normally high due to the number of users and business areas that may be affected. Inherent risk is the risk level or exposure without taking into account the actions that management has taken or might take.
C. Compliance risk is the penalty applied to current and future earnings for nonconformance to laws and regulations and may not be impacted by the number of users and business areas affected.
D. Residual risk is the remaining risk after management has implemented a risk response and is not based on the number of users or business areas affected.

R1-76 To be effective, risk management should be applied to:

A. those elements identified by a risk assessment.
B. any area that exceeds acceptable risk levels.
C. all organizational activities.
D. only those areas that have potential impact.

C is the correct answer.

Justification:

A. The elements of unacceptable risk will require treatment, but all activities are subject to risk management oversight. The assessing of risk, determining which risk is acceptable and which has the potential for impact are all functions of risk management.
B. Risk management must be holistic and should not be limited to areas that exceed acceptable risk levels; areas that are within acceptable risk levels may provide opportunity to improve performance by reducing control measures or assuming more risk.
C. While not all organizational activities will pose an unacceptable risk, the practice of risk management is ideally applied to all organizational activities.
D. When assessing risk, determining which risk is acceptable, which exceeds acceptable levels and which has the potential for impact are all functions of risk management.

R1-77 Corporate information security policy development should **PRIMARILY** be based on:

A. vulnerabilities.
B. threats.
C. assets.
D. impacts.

C is the correct answer.

Justification:

A. Absent a threat, vulnerabilities do not pose a risk. A vulnerability is defined as a weakness in the design, implementation, operation or internal control of a process that could expose the system to adverse threats from threat events.
B. A threat is defined as anything (e.g., object, substance, human) that is capable of acting against an asset in a manner that can result in harm. The policy is not written to directly address a threat, but rather to address the protection of assets from threats.
C. The corporate information security policy is based on management's commitment to protecting the assets of the enterprise (and relevant information of its business partners) from the various threats, risk and exposures that could occur.
D. Impact is not an issue if no threat exists. The impact is generally quantified as a direct financial loss in the short term or an ultimate (indirect) financial loss in the long term. Impact does not drive the development of the policy, but is a component of the policy.

R1-78 Which of the following vulnerabilities is the **MOST** serious and allows attackers access to data through a web application?

A. Validation checks are missing in data input fields.
B. Password rules do not enforce sufficient complexity.
C. Application transaction log management is weak.
D. The application and database share a single access ID.

A is the correct answer.

Justification:

A. When validation checks are missing in data input fields, attackers are able to exploit other weaknesses in the system. For example, they can submit a part of a structured query language (SQL) statement (SQL injection attack) to illegally retrieve application data, deface or even disable the web application. Input validation checks are effective countermeasures.

B. Noncomplex passwords may make accounts vulnerable to brute force attacks, but these can be countered in other ways besides complexity (e.g., lockout thresholds).

C. If application transaction log management is weak, there is a chance that confidential information is inadvertently written to the application transaction log. Sufficient care should therefore be given to log management. However, it is uncommon for attackers to use the log server to steal database information.

D. It is quite common that the application and database share a single access ID. If the supporting domain architecture is sufficiently secure, the overall risk is low.

R1-79 Which of the following combinations of factors helps quantify risk?

A. Probability and consequence
B. Impact and threat
C. Threat and exposure
D. Sensitivity and exposure

A is the correct answer.

Justification:

A. The quantification of risk is based on the probability (likelihood) of a threat exploiting a vulnerability resulting in a consequence (impact) of damage to an asset.

B. A threat is anything (e.g., object, substance, human) that is capable of acting against an asset in a manner that can result in harm. The impact is the effect of the threat on the asset. Threat and impact are not sufficient to quantify risk.

C. A threat is anything (e.g., object, substance, human) that is capable of acting against an asset in a manner that can result in harm. Exposure is the potential loss to an area due to the occurrence of an adverse event. Threat and exposure are not sufficient to quantify risk.

D. Sensitivity is a measure of the impact that improper disclosure of information may have on an enterprise. Exposure is the potential loss to an area due to the occurrence of an adverse event, but is not used to quantify risk.

R1-80 Which of the following requirements **MUST** be met during the initial stages of developing a risk management program?

A. Management acceptance and support have been obtained.
B. Information security policies and standards are established.
C. A management committee to provide program oversight exists.
D. The context and purpose of the program is defined.

D is the correct answer.

Justification:

A. Although an important component in the development of any managed program, obtaining management acceptance and support ideally occurs well before the development of the program.
B. Information security policies and standards are based on the decisions made in the planning phase of the program and are developed based on the outcomes and business objectives established by the business.
C. Management oversight of the risk management program is a monitoring control that is developed to ensure that the program meets business objectives. This process is established at the later stages of development, after the purpose of the program and the mechanics of its deployment have been established.
D. Initial requirements to determine the enterprise's purpose for creating an information security risk management program include determining the desired outcomes and defining objectives.

R1-81 The likelihood of an attack being launched against an enterprise is **MOST** dependent on:

A. the skill and motivation of the potential attacker.
B. the frequency that monitoring systems are reviewed.
C. the ability to respond quickly to any incident.
D. the effectiveness of the controls.

A is the correct answer.

Justification:

A. Factors that affect likelihood include the skill and motivation of the attacker as well as the knowledge of vulnerabilities, the use of popular hardware or software, the value of the asset (leads to motivation) and environmental factors such as politics, activists and disgruntled employees or dissatisfied customers.
B. Monitoring systems may detect an attack, but will not usually affect the likelihood of an attack. An exception to this is when the attacker knows that he/she is being monitored and is likely to be caught and then is less likely to launch an attack.
C. The ability to respond is important, but is only relevant once an attack has been conducted. It will not affect likelihood.
D. The controls may deter, prevent, detect or recover from an attack, but they will not necessarily affect the likelihood of someone trying to attack.

R1-82 Which of the following choices is the **MOST** important part of any outsourcing contract?

A. The right to audit the outsourcing provider
B. Provisions to assess the compliance of the provider
C. Procedures for dealing with incident notification
D. Requirements to encrypt hosted data

B is the correct answer.

Justification:
A. The service provider may not allow the outsourcing company the ability to audit them directly, but may provide a proof of compliance conducted by an independent auditor.
B. If the provision to monitor and hold a supplier accountable for security is not in the contract, then the outsourcing enterprise has no way to ensure compliance or proper handling of their data.
C. The outsourcing contract will not usually contain details on the procedures to follow when dealing with incidents.
D. There may not be a requirement to encrypt all data. Only sensitive data may require encryption.

R1-83 The **MOST** important external factors that should be considered in a risk assessment effort are:

A. proposed new security tools and technologies.
B. the number of viruses and other malware being developed.
C. international crime statistics and political unrest.
D. supply chain and market conditions.

D is the correct answer.

Justification:
A. It is always good to watch for new technologies and tools that can help the enterprise, especially ones that staff may want to bring into the office. But a risk assessment should not be based on proposed new products.
B. The number of new malware types being developed is something worth watching, but it is not a factor that the risk professional can use in the calculation of risk for a risk assessment report.
C. International crime statistics and political unrest may cause problems, but these are not the most important factors to be considered in a risk assessment effort.
D. Risk assessment should consider both internal and external factors, including supply chain and market conditions. Supply chain problems (e.g., lack of raw material, strikes at a transportation company or supplier) can severely interrupt operations. A new competitor in the market or even a new company opening up in the area may affect availability of trained staff or pose a risk to growth and profitability.

R1-84 The sales manager of a home improvement enterprise wants to expand the services available on the enterprise's web page to include sending free promotional samples of their products to prospective clients. What is the **GREATEST** concern the risk professional would have?

A. Are there any data privacy concerns about storing client data?
B. Are there any concerns about protecting credit card or payment data?
C. Can the system be misused by a person to obtain multiple samples?
D. Will the web site be able to handle the expected volume of traffic?

A is the correct answer.

Justification:

A. If the enterprise is sending a home improvement product to a client, the client data that have been entered on the web site may need to be protected against theft and disclosure.
B. Because this system offers free samples, there is no requirement to collect payment card data.
C. Obtaining multiple samples is primarily the concern of the sales department. The risk professional will point this out, but this is not the professional's greatest concern.
D. Handling traffic volume is a concern, but not the greatest concern in this scenario.

R1-85 Senior management will **MOST** likely have the highest tolerance for moving which of the following to a public cloud?

A. Credit card processing
B. Research and development
C. The legacy financial system
D. The corporate email system

D is the correct answer.

Justification:

A. Credit card processing can be a contender for public cloud computing, but in comparison to an email system, enforcing security requirements may be more challenging.
B. Research and development generally contains confidential information and is less likely than email to be outsourced to a cloud environment.
C. The legacy financial system contains not only financial information, but also will most likely be more complex to implement than an email system.
D. Considerations for moving processes and information to the cloud (public or hybrid) should include, among other factors, the criticality and complexity as well as the classification of the data supported by the process. Of the options offered, the corporate email system has the least competitive distinction, complexity and sensitive/highly classified information.

R1-86 Which of the following items is **MOST** important to consider in relation to a risk profile?

A. A summary of regional loss events
B. Aggregated risk to the enterprise
C. A description of critical risk
D. An analysis of historical loss events

B is the correct answer.

Justification:

A. The risk profile will consider regional events that could impact the enterprise, and will also consider systemic and other risk.
B. The risk profile is based on the aggregated risk to the enterprise, including historical risk, critical risk and emerging risk.
C. The risk profile will consider all risk, not just critical risk.
D. Analysis of historical loss events can assist in business continuity planning and risk assessment, but is incomplete for a risk profile.

R1-87 Which of the following factors determines the acceptable level of residual risk in an enterprise?

A. Management discretion
B. Regulatory requirements
C. Risk assessment results
D. Internal audit findings

A is the correct answer.

Justification:

A. Deciding what level of risk is acceptable to an enterprise is fundamentally a function of management. At its discretion, enterprise management may decide to accept risk. The target risk level for a control is therefore ultimately subject to management discretion.
B. Failure to comply with regulatory requirements has consequences, but those consequences are considered in the context of organizational risk. In some cases, the cost of failure to comply may be lower than the cost of compliance; in this case, management may decide to accept the risk.
C. The acceptable level of residual risk is determined by management and is not dependent on the results of the risk assessment.
D. The results of an internal audit are used to determine the actual level of residual risk for specific audit scope, but whether this level is acceptable is fundamentally a function of management.

R1-88 Which of the following environments typically represents the **GREATEST** risk to organizational security?

A. An enterprise data warehouse
B. A load-balanced, web server cluster
C. A centrally managed data switch
D. A locally managed file server

D is the correct answer.

Justification:
A. Enterprise data warehouses are generally subject to close scrutiny, good change control practices and monitoring.
B. Load-balanced, web server clusters are generally subject to close scrutiny, good change control practices and monitoring.
C. Centrally managed data switches are generally subject to close scrutiny, good change control practices and monitoring.
D. A locally managed file server will be the least likely to conform to organizational security policies because it is generally subject to less oversight and monitoring.

R1-89 Overall business risk for a particular threat can be expressed as the:

A. magnitude of the impact should a threat source successfully exploit the vulnerability.
B. likelihood of a given threat source exploiting a given vulnerability.
C. product of the probability and magnitude of the impact if a threat exploits a vulnerability.
D. collective judgment of the risk assessment team.

C is the correct answer.

Justification:
A. The magnitude of the impact of a successful threat provides only one factor.
B. The likelihood alone of the impact of a successful threat provides only one factor.
C. The product of the probability and magnitude of the impact provides the best measure of the risk to an asset.
D. The judgment of the risk assessment team defines the risk on an arbitrary basis and is not suitable for a scientific risk management process.

R1-90 When developing risk scenarios for an enterprise, which of the following is the **BEST** approach?

A. The top-down approach for capital-intensive enterprises
B. The top-down approach because it achieves automatic buy-in
C. The bottom-up approach for unionized enterprises
D. The top-down and the bottom-up approach because they are complementary

D is the correct answer.

Justification:
A. Both risk scenario development approaches should be considered simultaneously, regardless of the industry.
B. Both risk scenario development approaches should be considered simultaneously, regardless of the risk appetite.
C. Both risk scenario development approaches should be considered simultaneously, regardless of the industry.
D. The top-down and bottom-up risk scenario development approaches are complementary and should be used simultaneously. In a top-down approach, one starts from the overall business objectives and performs an analysis of the most relevant and probable risk scenarios impacting the business objectives. In a bottom-down approach, a list of generic risk scenarios is used to define a set of more concrete and customized scenarios, applied to the individual enterprise's situation.

R1-91 Which of the following documents **BEST** identifies an enterprise's compliance risk and the corrective actions in progress to meet these regulatory requirements?

A. An internal audit report
B. A risk register
C. An external audit report
D. A risk assessment report

B is the correct answer.

Justification:

A. Audit reports track audit findings and their respective actions, but based on the audit scope, do not necessarily include compliance-oriented findings or their risk. They generally do not include corrective actions in progress.

B. A risk register provides a report of all current identified risk within an enterprise, including compliance risk, with the status of the corrective actions or exceptions that are associated with them.

C. External audit reports are generally more reliable than internal audit reports due to the increased independence of the external auditor. Similar to internal audit reports, they do not generally include all relevant compliance risk, but may focus on a single regulatory requirement at a time, such as privacy, the Occupational Safety and Health Administration (OSHA) in the US, the US Sarbanes-Oxley Act of 2002, etc. They generally do not include corrective actions in progress.

D. Risk assessment reports may include compliance risk, but often do not include insights into the corrective actions that are ongoing or planned.

DOMAIN 2—IT RISK ASSESSMENT (28%)

R2-1 Which of the following uses risk scenarios when estimating the likelihood and impact of significant risk to the organization?

A. An IT audit
B. A security gap analysis
C. A threat and vulnerability assessment
D. An IT security assessment

C is the correct answer.

Justification:

A. An IT audit typically uses technical evaluation tools or assessment methodologies to enumerate risk; generally, this is done for the purpose of prioritizing audit projects or for delineating the scope of an audit.
B. A security gap analysis typically uses technical evaluation tools or assessment methodologies to enumerate risk or areas of noncompliance, but does not utilize risk scenarios.
C. A threat and vulnerability assessment typically evaluates all elements of a business process for threats and vulnerabilities and identifies the likelihood of occurrence and the business impact if the threats were realized.
D. An IT security assessment typically uses technical evaluation tools or assessment methodologies to enumerate risk or areas of noncompliance, but does not utilize risk scenarios.

R2-2 Which of the following will have the **MOST** significant impact on standard information security governance models?

A. Number of employees
B. Cultural differences between physical locations
C. Complexity of the organizational structure
D. Evolving legislative requirements

C is the correct answer.

Justification:

A. The number of employees has less impact on information security governance models because well-defined process, technology and personnel components intermingle to provide the proper governance.
B. The distance between physical locations has less impact on information security governance models because well-defined process, technology and personnel components intermingle to provide the proper governance.
C. Information security governance models are highly dependent on the complexity of the organizational structure. Some of the elements that impact organizational structure are multiple business units and functions across the organization, leadership and lines of communication.
D. Being current with changing legislative requirements should not have a major impact once good governance models are in place; therefore, governance will help in effective management of the organization's ongoing compliance.

R2-3 Which of the following will produce comprehensive results when performing a qualitative risk analysis?

A. A vulnerability assessment
B. Scenarios with threats and impacts
C. The value of information assets
D. Estimated productivity losses

B is the correct answer.

Justification:

A. A vulnerability assessment itself provides a one-sided view unless it is linked to specific risk scenarios that help determine likelihood and impact.

B. Using a list of possible scenarios with threats and impacts will better frame the range of risk and facilitate a more informed discussion and decision.

C. The value of information assets is an important starting point when performing a qualitative risk analysis. However, value without consideration of realistic threats and determination of likelihood and impact is not sufficient for a risk analysis.

D. Estimated productivity losses may be an important input into the projected magnitude of an impact. However, this choice is insufficient on its own.

R2-4 Who should be accountable for the risk to an IT system that supports a critical business process?

A. IT management
B. Senior management
C. The risk management department
D. System users

B is the correct answer.

Justification:

A. IT management is responsible for managing information systems on behalf of the business owners; they are not accountable for the risk.

B. The accountable party is senior management. While they may not be responsible for executing the risk management program, they are ultimately liable for the acceptance and mitigation of all risk.

C. The risk management department is responsible for the execution of the risk management program and will identify, evaluate and report on risk and risk response efforts; the department is not accountable for the risk.

D. System users are responsible for using the system properly and following procedures; they are not accountable for the risk.

R2-5 Which of the following is the **MAIN** outcome of a business impact analysis (BIA)?

A. Project prioritization
B. Criticality of business processes
C. The root cause of IT risk
D. Third-party vendor risk

B is the correct answer.

Justification:

A. Project prioritization is a core focus of program management with a focus on optimizing resource utilization; it is not the main outcome of a business impact analysis (BIA).
B. A BIA measures the total impact of tangible and intangible assets on business processes. Therefore, the sum of the value and opportunity lost as well as the investment and time required to recover is measured to determine the criticality of business processes.
C. A root cause analysis is a process of diagnosis to establish origins of events, which can be used to learn from consequences, typically from errors and problems, and is not an outcome of a BIA.
D. Third-party vendor risk should be documented during the BIA process, but it is not a main outcome.

R2-6 Which of the following provides the **MOST** valuable input to incident response efforts?

A. Qualitative analysis of threats
B. The annual loss expectancy (ALE) total
C. A vulnerability assessment
D. Penetration testing

A is the correct answer.

Justification:

A. Qualitative analysis of threats is an intuitive view of the outcome of various threat sources. Knowing the kinds of incidents that may occur in order of consequence will be of great benefit to incident response efforts.
B. The annual loss expectancy (ALE) total is the total cost associated with each source of risk and its probability of occurrence. This total may be of interest when preparing the budget, but cannot be directly linked to incident response efforts.
C. A vulnerability assessment is used to determine how easily security can be breached. This provides data about risk.
D. Penetration testing is used to provide tangible evidence that existing vulnerabilities can be exploited and the degree of difficulty to exploit them.

R2-7 Which of the following **BEST** describes the risk-related roles and responsibilities of an organizational business unit (BU)? The BU management team:

A. owns the mitigation plan for the risk belonging to their BU, while board members are responsible for identifying and assessing risk as well as reporting on that risk to the appropriate support functions.
B. owns the risk and is responsible for identifying, assessing and mitigating risk as well as reporting on that risk to the appropriate support functions and the board of directors.
C. carries out the respective risk-related responsibilities, but ultimate accountability for the day-to-day work of risk management and goal achievement belongs to the board members.
D. is ultimately accountable for the day-to-day work of risk management and goal achievement, and board members own the risk.

B is the correct answer.

Justification:

A. This choice is incorrect because the business unit (BU) management team owns both the risk management activities (identifying, assessing and reporting the mitigation plan for the risk belonging to their BU) and the reporting activities. The board members do not perform the risk identification, assessment and risk reporting functions.
B. This choice is the best statement because it assigns a senior management level owner to the risk and its resulting actions. Risk owners have the responsibility of identifying, measuring, monitoring, controlling and reporting on risk to executive management as established by the corporate risk framework.
C. This choice is incorrect because the ultimate accountability for the day-to-day work also belongs to the BU.
D. This choice is incorrect in the sense that it is reversed. The board members do not own the BU risk; the BU leader owns it, and along with the BU management team is accountable for the remediation efforts.

R2-8 Risk assessment techniques should be used by a risk practitioner to:

A. maximize the return on investment (ROI).
B. provide documentation for auditors and regulators.
C. justify the selection of risk mitigation strategies.
D. quantify the risk that would otherwise be subjective.

C is the correct answer.

Justification:

A. Maximizing the return on investment (ROI) may be a key objective for implementing risk responses, but is not part of the risk assessment process.
B. A risk assessment does not focus on auditors or regulators as primary recipients of the risk assessment documentation. However, risk assessment results may provide input into the audit process.
C. A risk practitioner should use risk assessment techniques to justify and implement a risk mitigation strategy as efficiently as possible.
D. Risk assessment is generally high-level, whereas risk analysis can be either quantitative or qualitative, based on the needs of the organization.

R2-9 Which of the following assessments of an enterprise's risk monitoring process will provide the **BEST** information about its alignment with industry-leading practices?

A. A capability assessment by an outside firm
B. A self-assessment of capabilities
C. An independent benchmark of capabilities
D. An internal audit review of capabilities

C is the correct answer.

Justification:

A. A capability assessment by an outside firm does not assess the enterprise against industry peers or competitors and only provides the opinion of the examiner as to what are/are not industry-leading practices.
B. A process capability self-assessment does not assess the enterprise against industry peers or competitors; it provides the opinion of the examiner, which in the case of a self-assessment is not even independent of the process to be reviewed.
C. An independent benchmark of capabilities allows an enterprise to understand its level of capability compared to other organizations within its industry. This allows the enterprise to identify industry-leading practices and its level of capability associated with these practices.
D. An internal audit review of capabilities does not assess the enterprise against industry peers or competitors; audits generally measure capabilities against corporate standards, not necessarily against industry-leading practices.

R2-10 Risk assessments should be repeated at regular intervals because:

A. omissions in earlier assessments can be addressed.
B. periodic assessments allow various methodologies.
C. business threats are constantly changing.
D. they help raise risk awareness among staff.

C is the correct answer.

Justification:

A. Omissions not found in earlier assessments do not justify regular reassessments.
B. This choice is incorrect because unless the environment changes, risk assessments should be performed using the same methodologies.
C. As business objectives and methods change, the nature and relevance of threats also change.
D. This choice is incorrect because there are better ways of raising security awareness than by performing a risk assessment, such as risk awareness training.

R2-11 Which of the following is **MOST** beneficial to the improvement of an enterprise's risk management process?

A. Key risk indicators (KRIs)
B. External benchmarking
C. The latest risk assessment
D. A maturity model

D is the correct answer.

Justification:

A. Key risk indicators (KRIs) are metrics that help monitor risk over time; they may be used to identify trends, but do not help define the desired state of the enterprise like a maturity model and thus are not the best option.
B. External benchmarking is useful to determine how other, similar enterprises manage risk, but does not help defined the desired state of the enterprise like a maturity model and thus is not the best option.
C. The latest risk assessment will be an input into the risk management process improvement effort, but does not help define the desired state of the enterprise like a maturity model and thus is not the best option.
D. A maturity model helps identify the *status quo* as well as the desired state and thus is most helpful when an enterprise desires to improve a business process, such as risk management.

R2-12 Which of the following is the **PRIMARY** reason for having the risk management process reviewed by independent risk auditors/assessors?

A. To ensure that the risk results are consistent
B. To ensure that the risk factors and risk profile are well defined
C. To correct any mistakes in risk assessment
D. To validate the control weaknesses for management reporting

B is the correct answer.

Justification:

A. Ensuring that risk results are consistent is very important to ensure that risk mitigation/management is effective and that is why risk management results are reviewed by independent risk auditors/assessors, who can be internal or external to the enterprise.
B. Risk profile and risk factors are defined during the risk assessment process; an independent review helps ensure that the underlying process is effective and helps identify areas for future improvement.
C. Risk assessment by an independent party is primarily performed to ensure and/or improve the quality of the risk assessment process, not to correct risk assessment mistakes.
D. The primary purpose of independent review is not to validate control weaknesses for management reporting, although it may be an outcome of the process.

R2-13 Which of the following provides the **GREATEST** support to a risk practitioner recommending encryption of corporate laptops and removable media as a risk mitigation measure?

A. Benchmarking with peers
B. Evaluating public reports on encryption algorithm in the public domain
C. Developing a business case
D. Scanning unencrypted systems for vulnerabilities

C is the correct answer.

Justification:
A. Benchmarking with peers does not help because peers will have a different risk environment and culture that cannot directly apply to one's own enterprise.
B. While evaluation of the solution in the public domain is important information, risk practitioners still need to analyze each solution in the context of their enterprise to provide the most valuable recommendation.
C. A business case has the business reasoning as to why the encryption solutions address the risk and also explains how the risk losses can be reduced.
D. Conducting a vulnerability assessment of unencrypted systems without proper business justification does not help.

R2-14 The **MOST** likely trigger for conducting a comprehensive risk assessment is changes to:

A. the asset inventory.
B. asset classification levels.
C. the business environment.
D. information security policies.

C is the correct answer.

Justification:
A. Additions and removals of assets from the asset inventory is an ongoing process and will not generally trigger a risk assessment.
B. Due to risk assessment one can understand the classification requirements, but this is not the main trigger for which the actual risk assessment is performed periodically.
C. Changes in the business environment in terms of new threats, vulnerabilities or changes to information assets deployment will act as a main trigger for conducting comprehensive risk assessment on a periodic basis. Based on periodic risk assessment, policies are modified rather than the other way around where risk assessment is performed based on policy changes already made.
D. Changes to information security policies may occur when a risk assessment indicates deficiencies at the security policies level; changes to security policies do not trigger risk assessments.

R2-15 Which of the following is used to determine whether unauthorized modifications were made to production programs?

A. An analytical review
B. Compliance testing
C. A system log analysis
D. A forensic analysis

B is the correct answer.

Justification:
A. Analytical review assesses the general control environment of an enterprise.
B. Compliance testing helps to verify that the change management process has been applied consistently.
C. It is unlikely that the system log analysis would provide information about the modification of programs.
D. Forensic analysis is a specialized technique for criminal investigation.

R2-16 An enterprise is hiring a consultant to help determine the maturity level of the risk management program. The **MOST** important element of the request for proposal (RFP) is the:

A. sample deliverable.
B. past experience of the engagement team.
C. methodology used in the assessment.
D. references from other organizations.

C is the correct answer.

Justification:
A. Sample deliverables only tell how the assessment is presented, not the process.
B. Past experience of the engagement team is not as important as the methodology used.
C. Methodology illustrates the process and formulates the basis to align expectations and the execution of the assessment. This also provides a picture of what is required of all parties involved in the assessment.
D. References from other organizations are important, but not as important as the methodology used in the assessment.

R2-17 The **BEST** time to perform a penetration test is after:

A. a high turnover in systems staff.
B. an attempted penetration has occurred.
C. various infrastructure changes are made.
D. an audit has reported control weaknesses.

C is the correct answer.

Justification:
A. Turnover in systems staff does not warrant a penetration test, although it may warrant a review of password change practices and configuration management.
B. Conducting a test after an attempted penetration is not as productive because an enterprise should not wait until it is attacked to test its defenses.
C. Changes in the systems infrastructure are most likely to inadvertently introduce new exposures.
D. Any exposure identified by an audit should be corrected before it would be appropriate to test.

R2-18 Which of the following should be in place before a black box penetration test begins?

A. A clearly stated definition of scope
B. Previous test results
C. Proper communication and awareness training
D. An incident response plan

A is the correct answer.

Justification:
A. A clearly stated definition of scope ensures a proper understanding of risk and success criteria.
B. Previous test results help define the scope.
C. Communication and awareness training are not a necessary requirement.
D. An incident response plan is not a necessary requirement. In fact, a penetration test could help promote the creation and execution of the incident response plan.

R2-19 Which of the following **BEST** assists a risk practitioner in measuring the existing level of development of risk management processes against their desired state?

A. A capability maturity model (CMM)
B. Risk management audit reports
C. A balanced scorecard (BSC)
D. Enterprise security architecture

A is the correct answer.

Justification:
A. The capability maturity model (CMM) grades processes on a scale of 0 to 5, based on their maturity, and is commonly used by entities to measure their existing state and then to determine the desired one.
B. Risk management audit reports offer a limited view of the current state of risk management.
C. A balanced scorecard (BSC) enables management to measure the implementation of strategy and assists in its translation into action.
D. Enterprise security architecture explains the security architecture of an entity in terms of business strategy, objectives, relationships, risk, constraints and enablers and provides a business-driven and business-focused view of security architecture.

R2-20 Which of the following is the **BEST** way to ensure that a corporate network is adequately secured against external attack?

A. Utilize an intrusion detection system (IDS).
B. Establish minimum security baselines.
C. Implement vendor recommended settings.
D. Perform periodic penetration testing.

D is the correct answer.

Justification:
A. An intrusion detection system (IDS) may detect an attempted attack, but it will not confirm whether the perimeter is secure.
B. Minimum security baselines are beneficial, but they will not provide the level of assurance that is provided by penetration testing.
C. Applying vendor recommended settings is beneficial, but it will not provide the level of assurance that is provided by penetration testing.
D. Penetration testing is the best way to ensure that perimeter security is adequate.

R2-21 A third party is engaged to develop a business application. Which of the following **BEST** measures for the existence of back doors?

A. Security code reviews for the entire application
B. System monitoring for traffic on network ports
C. Reverse engineering the application binaries
D. Running the application from a high-privileged account on a test system

A is the correct answer.

Justification:
A. Security code reviews for the entire application are the best measure and involve reviewing the entire source code to detect all instances of back doors.
B. System monitoring for traffic on network ports is not able to detect all instances of back doors, is time consuming and takes a lot of effort.
C. Reverse engineering the application binaries may not provide any definite clues.
D. Back doors do not surface by running the application on high-privileged accounts because back doors are usually hidden accounts in the applications.

R2-22 A substantive test to verify that tape library inventory records are accurate is:

A. determining whether bar code readers are installed.
B. conducting a physical count of the tape inventory.
C. checking whether receipts and issues of tapes are accurately recorded.
D. determining whether the movement of tapes is authorized.

B is the correct answer.

Justification:
A. Testing the existence of bar code readers is a compliance test, not a substantive test. A substantive test includes gathering evidence to evaluate the integrity of individual transactions, data or other information.
B. A substantive test includes gathering evidence to evaluate the integrity of individual transactions, data or other information. Conducting a physical count of the tape inventory is a substantive test.
C. Confirming that receipts and issues of tapes are accurately recorded is a compliance test, not a substantive test. A substantive test includes gathering evidence to evaluate the integrity of individual transactions, data or other information.
D. Testing the approval of tape movements is a compliance test, not a substantive test. A substantive test includes gathering evidence to evaluate the integrity of individual transactions, data or other information.

R2-23 The **BEST** method for detecting and monitoring a hacker's activities without exposing information assets to unnecessary risk is to utilize:

A. firewalls.
B. bastion hosts.
C. honeypots.
D. screened subnets.

C is the correct answer.

Justification:
A. Firewalls attempt to keep the hacker out, which is a preventive control.
B. Bastion hosts attempt to keep the hacker out, which is a preventive control.
C. The best choice for diverting a hacker away from critical files and alerting security of the hacker's presence are honeypots, often referred to as decoy files.
D. Screened subnets or demilitarized zones (DMZs) provide a middle ground between the trusted internal network and the external, untrusted Internet.

R2-24 Which of the following is the **BEST** way to verify that critical production servers are utilizing up-to-date antivirus signature files?

A. Check a sample of servers.
B. Verify the date that signature files were last pushed out.
C. Use a recently identified benign virus to test whether it is quarantined.
D. Research the most recent signature file, and compare it to the console.

A is the correct answer.

Justification:
A. The only effective way to check the currency of signature files is to look at a sample of servers.
B. The fact that an update was pushed out to a server does not guarantee that it was properly loaded onto that server. In conjunction with the sample testing, the process for updating the signature files should be verified.
C. Personnel should never release a virus, no matter how benign.
D. Comparing the vendor's most recent signature file to the management console is not indicative of whether the file was properly loaded on the server.

R2-25 Which of the following **BEST** helps identify information systems control deficiencies?

A. Gap analysis
B. The current IT risk profile
C. The IT controls framework
D. Countermeasure analysis

A is the correct answer.

Justification:
A. Controls are deployed to achieve the desired control objectives based on risk assessments and business requirements. The gap between desired control objectives and actual IS control design and operational effectiveness identifies IS control deficiencies.
B. Without knowing the gap between desired state and current state, one cannot identify the control deficiencies.
C. The IT controls framework is a generic document with no information such as desired state of IS controls and current state of the enterprise; therefore, it will not help in identifying IS control deficiencies.
D. Countermeasure analysis only helps in identifying deficiencies in countermeasures and not in the full set of primary controls.

R2-26 When assessing the performance of a critical application server, the **MOST** reliable assessment results may be obtained from:

A. activation of native database auditing.
B. documentation of performance objectives.
C. continuous monitoring.
D. documentation of security modules.

C is the correct answer.

Justification:
A. Native database audit logs are a good detective control, but do not provide information about the application server performance.
B. Documentation of performance objectives is important, but does not provide information about the application server performance.
C. It is essential to obtain monitoring data in a consistent manner to achieve reliable results. Changing the monitoring methodology frequently does not enable time-series data comparison.
D. Documentation of associated security modules may be helpful, but does not provide information about the application server performance.

R2-27 The **PRIMARY** goal of a postincident review is to:

A. gather evidence for subsequent legal action.
B. identify ways to improve the response process.
C. identify individuals who failed to take appropriate action.
D. make a determination as to the identity of the attacker.

B is the correct answer.

Justification:
A. Evidence should already have been gathered earlier in the process.
B. The goal of a postincident review is to identify ways to improve the incident response process.
C. A postincident review should not focus on finding and punishing individuals who did not take appropriate action, but rather on establishing a process to reduce the likelihood of similar incidents in the future and to improve the incident response process.
D. Identification of the attacker is not an objective of the postincident review process.

R2-28 IT risk is measured by its:

A. level of damage to IT systems.
B. impact on business operations.
C. cost of countermeasures.
D. annual loss expectancy (ALE).

B is the correct answer.

Justification:

A. Measurement by IT damage alone is not comprehensive enough; business risk must also be considered.
B. IT risk includes information and communication technology risk, but is primarily measured by its impact on the business. IT risk is the business risk associated with the use, ownership, operation, involvement, influence and adoption of IT within an enterprise.
C. The cost and benefit of countermeasures is concerned with risk response, not with risk assessment.
D. Annual loss expectancy (ALE) is a quantitative measure and must be used in conjunction with qualitative measures.

R2-29 Deriving the likelihood and impact of risk scenarios through statistical methods is **BEST** described as:

A. quantitative risk analysis.
B. risk scenario analysis.
C. qualitative risk analysis.
D. probabilistic risk assessment.

A is the correct answer.

Justification:

A. The essence of quantitative risk assessment is to derive the likelihood and impact of risk scenarios, based on statistical methods and data.
B. A risk scenario analysis generally includes several risk analysis methods, including quantitative, semiquantitative and qualitative. The question stem describes only the quantitative risk analysis method.
C. A qualitative risk analysis would use nonquantitative measures to estimate the likelihood and impact of adverse events; these might include low, medium and high for likelihood and low, medium, high and catastrophic for the impact.
D. Probabilistic risk assessments are mostly used for the assessment of risk related to complex engineered technology (e.g., nuclear plants, airplanes). They rely on a systematic and comprehensive methodology and consider both quantitative and qualitative risk analysis. The question stem describes only the quantitative risk analysis method.

R2-30 During an internal risk assessment in a global enterprise, a risk manager notes that local management has proactively mitigated some of the high-level risk related to the global purchasing process. This means that:

A. the local management is now responsible for the risk.
B. the risk owner is the corporate chief risk officer (CRO).
C. the risk owner is the local purchasing manager.
D. corporate management remains responsible for the risk.

D is the correct answer.

Justification:
A. While the local management has mitigated the risk, corporate management remains responsible for the risk.
B. The corporate chief risk officer (CRO) is responsible for the corporate risk management program, yet does not own the risk related to the global purchasing process.
C. The risk owner is the global purchasing manager.
D. Corporate management remains responsible for the risk, even when the risk response is executed at a lower organizational level.

R2-31 Which of the following **BEST** estimates the likelihood of significant events impacting an enterprise?

A. Threat analysis
B. Cost-benefit analysis
C. Scenario analysis
D. Countermeasure analysis

C is the correct answer.

Justification:
A. Threat analysis does not provide the complete picture to estimate likelihood. While there may be a threat, many other factors, including existing controls, must be considered to determine the likelihood of a threat.
B. Cost-benefit analysis is used for the selection of control and does not help estimate the likelihood of significant events.
C. Scenario analysis along with vulnerability analysis best helps determine whether a particular risk is relevant to the enterprise and helps estimate the likelihood of significant events impacting the enterprise.
D. Countermeasure analysis is used for analyzing those controls that address specific attacks, at times while the attack is occurring. Countermeasure analysis does not help estimate the likelihood of significant events.

R2-32 Which of the following is **MOST** important during the quantitative risk analysis process?

A. Statistical analysis
B. Decision trees
C. Expected monetary value (EMV)
D. Net present value (NPV)

C is the correct answer.

Justification:

A. Statistical analysis may be used because it helps risk managers make better decisions under conditions of uncertainty. However, it is not the most important.
B. Decision trees help determine the optimal course of action in complex situations with uncertain outcomes.
C. Expected monetary value (EMV) is the weighted average of probable outcomes. It represents the expected average payoff if you made that decision, using the same payoffs and probabilities, an infinite number of times.
D. Net present value (NPV) is calculated by using an after-tax discount rate of an investment and a series of expected incremental cash outflows (the initial investment and operational costs) and cash inflows (cost savings or revenues) that occur at regular periods during the life cycle of the investment.

R2-33 When would an enterprise project management department **PRIMARILY** use risk analysis?

A. During preparation for natural disasters
B. During go/no go decisions
C. During workplace safety training development
D. During regulation bulletin reviews

B is the correct answer.

Justification:

A. The business continuity department may use the risk analysis results to assist in planning for natural disasters, but the enterprise project management department would not.
B. The project management department can use the results of risk analysis to assist in making go/no go decisions at critical stages/phases of a project as well as during project planning to help account for potential threats and known vulnerabilities.
C. The facilities/safety department may use the risk analysis results to understand safety-related risk and may work with training to build safety training programs, but it is not the primary use.
D. The compliance department would use risk analysis to gain understanding of the potential impact to the enterprise based on a newly communicated regulation bulletin.

R2-34 The **PRIMARY** reason to have the risk management process reviewed by independent risk management professional(s) is to:

A. validate cost-effective solutions for mitigating risk.
B. validate control weaknesses detected by the internal team.
C. assess the validity of the end-to-end process.
D. assess that the risk profile and risk factors are properly defined.

C is the correct answer.

Justification:

A. This is not necessary because cost-effective solutions cannot be provided by the internal teams.
B. The internal team can find weaknesses. It is not necessary to involve external risk professionals to validate the weaknesses as detected by the internal team.
C. Because independent risk professionals are not affected by the internal politics and other factors, they would provide an unbiased assessment on the validity of the end-to-end risk management process.
D. The risk profile and risk factors are properly defined when the risk assessment process is performed correctly. An independent assessment may result in further improvements.

R2-35 Which of the following is **BEST** suited for the review of IT risk analysis results before the results are sent to management for approval and use in decision making?

A. An internal audit review
B. A peer review
C. A compliance review
D. A risk policy review

B is the correct answer.

Justification:

A. An internal audit review is not best suited for the review of IT risk analysis results. Internal auditing is an independent, objective assurance and consulting activity designed to add value and improve an enterprise's operations. It helps an organization accomplish its objectives by bringing a systematic, disciplined approach to evaluate and improve the effectiveness of risk management, control and governance processes.
B. It is effective, efficient and good practice to perform a peer review of IT risk analysis results before sending them to management.
C. A compliance review is not best suited for the review of IT risk analysis results. Compliance reviews measure the conformance with a specific, measurable standard.
D. A review of the risk policy will change the contents and methods of the risk analysis eventually, but this is not a way of reviewing IT risk analysis results before sending them to management.

R2-36 Which of the following is responsible for evaluating the effectiveness of existing internal information security (IS) controls within an enterprise?

A. The data owner
B. Senior management
C. End users
D. The system auditor

D is the correct answer.

Justification:
A. The data owner defines the business requirements for internal IS controls.
B. Senior management is accountable to ensure that existing internal IS controls are effective.
C. End users are not responsible for evaluating the effectiveness of existing internal IS controls.
D. The system auditor is responsible for providing continuous feedback to senior management about the effectiveness of internal controls within the enterprise. This is part of his/her normal routine responsibilities.

R2-37 Which of the following **BEST** enables a peer review of an enterprise's risk management process?

A. A balanced scorecard (BSC)
B. An industry survey
C. A capability maturity model (CMM)
D. A framework

C is the correct answer.

Justification:
A. A balanced scorecard (BSC) is a coherent set of performance measures organized into four categories that includes traditional financial measures, but adds customer, internal business process, and learning and growth perspectives.
B. An industry survey does provide a view of current practices. Because survey results are generally presented in an aggregated manner, they do not enable a peer review of an enterprise's risk management process.
C. A capability maturity model (CMM) contains the essential elements of effective processes for one or more disciplines. It also describes an evolutionary improvement path from *ad hoc*, immature processes to disciplined, mature processes with improved quality and effectiveness.
D. A framework is a set of concepts, assumptions and practices that define how something can be approached or understood, the relationships among the entities involved, the roles of those involved, and the boundaries.

R2-38 Which of the following **BEST** ensures the overall effectiveness of a risk management program?

A. Obtaining feedback from all end users
B. Assigning a dedicated risk manager to run the program
C. Applying quantitative risk methodologies
D. Participating relevant stakeholders

D is the correct answer.

Justification:
A. It is not feasible to obtain feedback from all end users even though that would give the most complete view of an enterprise's risk universe. This is because each employee potentially has a unique perspective to his/her own sphere of control.
B. Assigning a dedicated risk manager, such as a program manager, is a good option, but is less effective without stakeholder involvement.
C. Either methodology can be selected to create an effective risk management process. Selection of a quantitative or qualitative risk assessment methodology depends on the needs of the organization.
D. Without participation (e.g., supervision, risk monitoring, risk-related decisions) from stakeholders of the enterprise, the risk management program is ineffective because stakeholders are the critical parties making risk-related decisions or directly/indirectly impacting the risk response.

R2-39 Which of the following **BEST** ensures that identified risk is kept at an acceptable level?

A. Reviewing of the controls periodically, according to the risk action plan
B. Listing each risk as a separate entry in the risk register
C. Creating a separate risk register for every department
D. Maintaining a key risk indicator (KRI) for assets in the risk register

A is the correct answer.

Justification:
A. Controls deployed according to the risk action plan to manage the risk should provide the desired results because the risk action plan is based on management acceptance of residual risk and on the steps in the risk action plan.
B. Listing each risk as a separate entry in the risk register may help in better evaluating the risk, but the register in itself does not ensure risk management of identified risk at a reasonable level.
C. Creating a separate risk register for every department may help in a better risk assessment exercise, but the register in itself does not ensure risk management of identified risk at a reasonable level.
D. Maintaining a key risk indicator (KRI) for assets in the risk register may improve the overall risk management cycle, but the register in itself does not ensure that the management of identified risk has been performed according to the risk action plan.

R2-40 What is the **FIRST** step for a risk practitioner when an enterprise has decided to outsource all IT services and support to a third party?

A. Validate that the internal systems of the service provider are secure.
B. Enforce the regulations and standards associated with outsourcing data management for restrictions on transborder data flow.
C. Ensure that security requirements are addressed in all contracts and agreements.
D. Build a business case to perform an onsite audit of the third-party vendor.

C is the correct answer.

Justification:

A. A risk practitioner will rarely have access to validate the security of a third party. He/she must seek other assurances from an external audit or other standards.
B. A risk practitioner can advise on risk associated with outsourcing and regulations, but cannot enforce such rules.
C. A contract only covers the topics listed in the contract. If security is not explicitly included in the contract terms, the enterprise may not be properly protected.
D. Even though IT management has been outsourced, the enterprise that outsourced the service function remains responsible for protecting its data.

R2-41 Which of the following capability dimensions is **MOST** important when using a maturity model for assessing the risk management process?

A. Effectiveness
B. Efficiency
C. Profitability
D. Performance

D is the correct answer.

Justification:

A. Effectiveness is a subset of the performance capability criterion.
B. Efficiency is a subset of the performance capability criterion.
C. Profitability is generally not considered when using a capability maturity model for assessing the risk management process.
D. The most important criterion when using a capability maturity model is performance. Performance is achieved when the implemented process achieves its purpose and is thus the most important capability dimension when using a capability maturity model for assessing the risk management process.

R2-42 An enterprise's corporate policy specifies that only failed and successful access attempts are logged. What is the **PRIMARY** risk to the enterprise?

A. The source IP address is not logged.
B. The destination IP address is not logged.
C. Login information can be lost if the data are not automatically moved to secondary storage.
D. The details of what commands were executed is missing.

D is the correct answer.

Justification:
A. Source IP addresses are logged for failed and successful access attempts.
B. The destination IP address is already known.
C. While servers have limited storage, backups are generally stored using first in, first out (FIFO); so, unless secondary storage is set up to move these data in a timely fashion, logging information can be lost. This is a risk, but not a primary one.
D. Login statistics alone do not provide adequate forensic information. Logs need to contain the commands executed (transactions invoked) to know what activities were conducted once someone successfully logs in because that may be someone who successfully hacked into the network.

R2-43 During an internal assessment, an enterprise notes that only a couple dozen hard-coded individual transactions are being logged, which does not encompass what should be logged to meet regulatory requirements. The individual server log files use first in, first out (FIFO). Most files recycle in less than 24 hours. What is the **MOST** financially damaging vulnerability associated with the current logging practice?

A. The log data stored recycles in less than 24 hours.
B. The log files are stored on the originating servers.
C. Regulation-related transactions may not be tracked.
D. Transactions being logged are hard coded.

C is the correct answer.

Justification:
A. Log data recycling in less than 24 hours can impair the ability to perform root cause analysis, but is generally not as financially damaging as failing to properly track regulation-related transactions.
B. Backing up log files to the same server can have a significant impact because in the case of an incident, log files will also be affected. Additionally, privileged accounts can make changes and modify logged data. However, this practice is generally not as financially damaging as failing to properly track regulation-related transactions.
C. The enterprise may be fined for failing to properly track regulation-related transactions.
D. The scope of what is being logged is limited because only those transactions that are explicitly stated to be included will show activities. This means that the majority of transactions are being processed without an audit trail.

R2-44 Testing the compliance of a response and recovery plan should begin with conducting a:

A. tabletop exercise.
B. review of archived logs.
C. penetration test.
D. business impact analysis (BIA).

A is the correct answer.

Justification:

A. Tabletop exercises are simulated scenarios designed to test the response capability of an enterprise to a given event. The scenarios require a coordinated response to a realistic situation that develops in real time with participants gathered to formulate responses to each development. Tabletop exercises have been used extensively in the police, fire and emergency medical services (EMS) fields as a method for gathering key personnel to practice response to, and recovery from, a wide range of incidents that can occur within a specific jurisdiction.

B. Logs provide a way to trace the activities performed during the vulnerability assessment.

C. Penetration tests highlight specific weaknesses; but while these tests generally are very controlled, they do not provide the depth and breadth of a tabletop exercise.

D. A business impact analysis (BIA) provides the "as of" view based on the process owner and is input to the response and recovery plan; it should be used as the basis for building the test scripts to validate compliance, but in and of itself, it is not a testing tool.

R2-45 The IT department wants to use a server for an enterprise database, but the server hardware is not certified by the operating system (OS) or the database vendor. A risk practitioner determines that the use of the database presents:

A. a minimal level of risk.
B. an unknown level of risk.
C. a medium level of risk.
D. a high level of risk.

B is the correct answer.

Justification:

A. If this were true, most hardware certifications and support departments would vanish! While it is true that there are standard interfaces on most PCs and servers (e.g., USB ports, SATA, HDMI), the internal architecture and BIOS calls of all PCs and servers are different from vendor to vendor, which means that unless the hardware is certified to work with at least the OS (or both the OS and the database), you are in for a support nightmare.

B. Because the hardware is not certified by its manufacturer to work without major issues using the OS or the database software, the risk is unknown. An enterprise database is a critical application and no one would approve an unknown risk like this one.

C. Using unknown hardware for an enterprise database system is an unknown risk and is usually such a high risk that no enterprise would try this with hardware that is not certified because the downtime and support are almost always higher in the long term than the purchase price of the hardware.

D. Because database vendors do not support hardware directly, this answer is incorrect. The database vendor supports and works with different OSs. The OS vendor supports and works with the hardware or its vendor.

R2-46 The **PRIMARY** benefit of using a maturity model to assess the enterprise's data management process is that it:

A. can be used for benchmarking.
B. helps identify gaps.
C. provides goals and objectives.
D. enforces continuous improvement.

B is the correct answer.

Justification:
A. While maturity models can be used for benchmarking, the benchmarking is not a primary benefit.
B. Maturity models identify gaps between the current and the desired state to help enterprises determine necessary remediation efforts.
C. While maturity models help determine goals and objectives, the primary value is that they identify the current state as well as the desired state. The gaps between the two help define areas where action needs to be taken.
D. Continuous improvement may not be the objective of an enterprise, particularly when the current maturity level meets its needs.

R2-47 What do different risk scenarios on the same bands/curve on a risk map indicate?

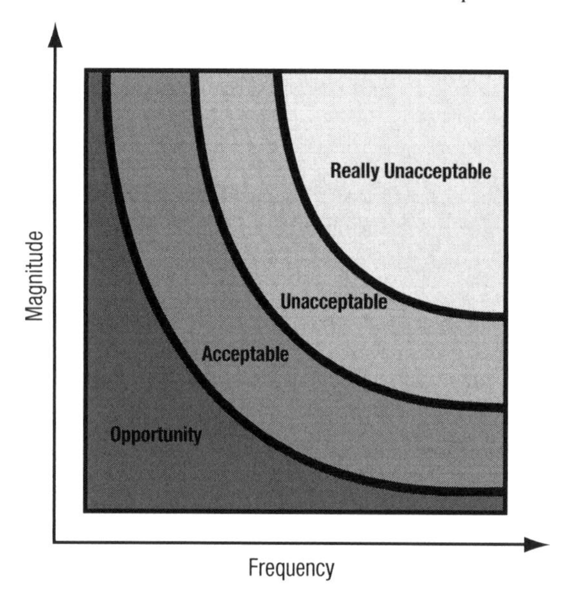

A. All risk scenarios on the same curve of a risk map have the same level of risk.
B. All risk scenarios on the same curve of a risk map have the same magnitude of impact.
C. All risk scenarios on the same curve of a risk map require the same risk response.
D. All risk scenarios on the same curve of a risk map are of the same type.

A is the correct answer.

Justification:
A. Each curve on a risk map indicates risk scenarios with the same level of risk: While one risk may have a frequency of 10,000 and an impact of $15/occurrence, another risk may have a frequency of 1 and an impact of $150,000/occurrence. Risk is derived as a function of magnitude multiplied by frequency.
B. The magnitude of impact for each risk scenario on a curve increases as you move left on the curve and up along the Y axis.
C. Risk scenarios on the same curve of a risk map do not necessarily require the same risk response. While all risk scenarios on the same curve have the same level of risk, the actual risk response may differ and be influenced by factors, such as cost of remediation.
D. The risk scenarios on the same curve of a risk map have the same level of risk, but are not necessarily of the same type. Risk scenarios with the same level of risk may vary greatly in type.

R2-48 The goal of IT risk analysis is to:

A. enable the alignment of IT risk management with enterprise risk management (ERM).
B. enable the prioritization of risk responses.
C. satisfy legal and regulatory compliance requirements.
D. identify known threats and vulnerabilities to information assets.

B is the correct answer.

Justification:

A. Aligning IT risk management with enterprise risk management (ERM) is important to ensure the cost-effectiveness of the overall risk management process. However, risk analysis does not enable such an alignment.
B. Risk analysis is a process by which likelihood and magnitude of IT risk scenarios are estimated. Risk analysis is conducted to ensure that areas with greatest risk likelihood and impact are prioritized above those areas with lower likelihood and impact. Prioritization of IT risk helps maximize return on investment (ROI) on risk responses.
C. Risk analysis evaluates risk on the basis of likelihood and impact and includes financial, environmental, regulatory and other risk. It does look at regulatory risk as one of many types of risk that the enterprise faces, and is not specifically designed to satisfy legal and regulatory compliance requirements.
D. Risk analysis occurs after risk identification and evaluation. Risk identification does identify known threats and vulnerabilities. Risk evaluation assesses the risk and creates valid risk scenarios. Risk analysis quantifies risk along the vectors of likelihood and impact to facilitate the prioritization of risk responses.

R2-49 Which of the following factors should be assessed after the likelihood of a loss event has been determined?

A. Magnitude of impact
B. Risk tolerance
C. Residual risk
D. Compensating controls

B is the correct answer.

Justification:

A. Risk tolerance is the acceptable deviation from acceptable risk. This is taken into account once risk has been quantified, which is dependent on determining the magnitude of impact.
B. Once likelihood has been determined, the next step is to determine the magnitude of impact.
C. Residual risk is the remaining risk after management has implemented a risk response. This cannot be calculated until the controls have been selected.
D. Compensating controls are internal controls that reduce the risk of an existing or potential control weakness resulting in errors and omissions. They would not be assessed directly in conjunction with assessing the likelihood of a loss event.

R2-50 Which of the following approaches **BEST** helps an enterprise achieve risk-based organizational objectives?

A. Ensure that asset owners perform annual risk assessments.
B. Review and update the risk register regularly.
C. Assign a steering committee to the risk management process.
D. Embed risk management activities into business processes.

D is the correct answer.

Justification:
A. Performing a risk assessment does not achieve risk-based organizational objectives.
B. Maintaining a risk register may be good for identifying issues, but does not achieve risk-based organizational objectives.
C. Assigning a steering committee to the risk management process will aid in alignment; however, it will not be as effective as embedding risk management activities into business processes.
D. The primary objective of embedding risk management activities into business processes is to achieve risk-based organizational objectives in the most effective manner possible.

R2-51 What is the **ULTIMATE** goal of risk aggregation?

A. To prevent attacks from exploiting a combination of low-level types of risk that individually have not been properly mitigated
B. To address the threat of an exploit that attacks a system through a series of individual attacks
C. To ensure that the combined value of low-level risk is not overlooked in the risk management process
D. To stop attackers from gaining low-level access and then escalating their attack through access aggregation

C is the correct answer.

Justification:
A. Risk aggregation is the process of integrating risk assessments at a corporate level to obtain a complete view on the overall risk for the enterprise and assessing more specifically for exploit opportunities, whereas a series or combination of types of low-level risk left unaddressed can provide an attacker a means to exploit one or more enterprise resources.
B. An exploit using several different attacks in sequence to launch an attack is not an example of risk aggregation, but of a chained exploit.
C. Individual risk that may pose the threat of minimal impact on its own may constitute a significant overall risk if several instances of risk can be aggregated and work together to defeat the risk controls. For example, one machine is not able to effectively flood a network, but many machines working together can. Another example is a database that contains many pieces of information. If those individual pieces of information can be aggregated together, the attacker may learn information classified at a higher level.
D. Many attacks today start at a low level and then increase their capability as they move through the system. Such attacks are often facilitated through accounts with excessive levels of access. This is not an example of risk aggregation.

R2-52 How can a risk professional calculate the total impact to operations if hard drives supporting a critical financial system fail?

A. Calculate the replacement cost for failed equipment and the time needed for service restoration.
B. Gather the cost estimates from the finance department to determine the cost.
C. Use quantitative and qualitative methods to examine the effect on all affected business areas.
D. Review regulatory and contractual requirements to quantify liabilities.

C is the correct answer.

Justification:

A. The risk is not solely dependent on the IT-related costs of the failed equipment. The impact on the business must also be determined.
B. Gathering cost estimates is a quantitative method of risk assessment, and may not be a reflection of the total impact of the event if only the finance department's costs are taken into consideration.
C. An event in one department may affect many areas of the enterprise, and the impact on all areas should be included in the risk calculation. Using quantitative and qualitative methods will provide the information needed to examine the effects of the failure.
D. The regulatory and contractual requirements must be included in the risk calculation, but they are not the only relevant factors.

R2-53 When assessing the capability of the risk management process, a regulatory body would place the **GREATEST** reliance on:

A. a peer review.
B. an internal review.
C. an external review.
D. a process capability review.

C is the correct answer.

Justification:

A. A peer review is a means to evaluate the work of another member of a team. The level of independence is not as high as that of an external review.
B. An internal review may be subject to management influence and does not have the same level of independence as an external review.
C. Regulatory entities generally utilize assessments that are performed by an objective and independent third party. Of the choices presented, an external review is the most objective and independent.
D. A process capability review determines the capability of a process, such as the risk management process. However, the option does not indicate the level of independence and objectivity and is thus not the best option.

R2-54 Once a risk assessment has been completed, the documented test results should be:

A. destroyed.
B. retained.
C. summarized.
D. published.

B is the correct answer.

Justification:

A. Test results should be stored in a secure manner for future reference and comparison.
B. Test results should be retained in order to ensure that future tests can be compared with past results and ensure reporting consistency.
C. Test results are summarized as part of the risk assessment process.
D. Assessment results are not usually published due to vulnerability disclosure.

R2-55 A company is confident about the state of its organizational security and compliance program. Many improvements have been made since the last security review was conducted one year ago. What should the company do to evaluate its current risk profile?

A. Review previous findings and ensure that all issues have been resolved.
B. Conduct follow-up audits in areas that were found deficient in the previous review.
C. Monitor the results of the key risk indicators (KRIs) and use those to develop targeted assessments.
D. Perform a new enterprise risk assessment using an independent expert.

D is the correct answer.

Justification:

A. Even though the findings of the previous test have been addressed, a new risk assessment would be the best way to indicate the effectiveness of the controls and uncover any new risk.
B. Making changes in one area may cause inadvertent effects in other areas. Therefore, an enterprisewide risk assessment would be better than just testing the previous areas.
C. Monitoring key risk indicators (KRIs) can indicate areas of emerging risk and unsatisfactory security levels, and this should drive individual tests during the year. However, KRIs are one of many tools used to determine the entire enterprise risk profile. One common mistake when implementing KRIs is selecting too many.
D. The best way to ensure that an enterprise's security posture is still within compliance is to conduct another risk assessment. It has been a year, and a lot can change in a year. Using an independent expert can provide more objective results than using an internal person who would be testing his/her own work.

R2-56 What is a **PRIMARY** advantage of performing a risk assessment on a consistent basis?

A. It lowers the costs of assessing risk.
B. It provides evidence of threats.
C. It indicates trends in the risk profile.
D. It eliminates the need for periodic audits.

C is the correct answer.

Justification:
A. There may be some minor cost benefits to performing risk assessments on a consistent basis, but that is not the main benefit.
B. A risk assessment provides evidence of risk; however, it is not intended to provide evidence of threats.
C. Tracking trends in evolving risk is of significant benefit to managing risk and ensuring that appropriate controls are in place.
D. The performance of risk assessment on a consistent basis does not preclude the requirement to perform periodic independent audits.

R2-57 The capability maturity model (CMM) is based on:

A. the training of staff to ensure consistent knowledge transfer.
B. the development of new controls to replace aging or diminished controls.
C. the application of standard, repeatable processes that can be measured.
D. users developing new innovative solutions to problems.

C is the correct answer.

Justification:
A. Training staff is the transfer of knowledge. Capability maturity models (CMMs) address the consistent application of procedures, not training.
B. A CMM relies on consistently applied metrics, not the replacement of controls.
C. The use of maturity models is based on the development, use and measuring of a consistent set of procedures and activities. A maturity model contains the essential elements of effective processes for one or more disciplines. It also describes an evolutionary improvement path from *ad hoc*, immature processes to disciplined, mature processes with improved quality and effectiveness.
D. Empowering users to develop new solutions applies to total quality management (TQM), while CMMs encourage the use of standard, repeatable procedures.

R2-58 The **MOST** important task in system control verification is:

A. monitoring password resets.
B. detecting malware.
C. managing alerts.
D. performing log reviews.

C is the correct answer.

Justification:
A. Password reset monitoring by itself does not allow for verification that system controls are working properly.
B. Malware monitoring by itself does not allow for verification that system controls are working properly.
C. The most important task in system control verification is managing the response time to critical alerts and alarms.
D. Log reviews are a fundamental system control verification process because they allow verification of how the system and its settings are performing; however, reviews are not as critical as responding to alerts.

R2-59 Which automated monitoring technique in an application uses triggers to indicate a suspicious condition?

A. Snapshots
B. An integrated test facility
C. Monitor hooks
D. Continuous and intermittent simulation

C is the correct answer.

Justification:
A. The snapshots technique takes a picture of a system status to identify specific values or configuration settings.
B. An integrated test facility feeds dummy transactions into the production flow and compares them to predetermined results.
C. The monitor hooks technique has embedded hooks in the application that act as triggers if certain conditions are met.
D. In continuous and intermittent simulation, data are monitored only if they meet certain criteria.

R2-60 Which of the following is the **BEST** reason to perform a risk assessment?

A. To satisfy regulatory requirements
B. To budget appropriately for needed controls
C. To analyze the effect on the business
D. To help determine the current state of risk

D is the correct answer.

Justification:
A. Performing a risk assessment may satisfy regulatory requirements, but is not the reason to perform a risk assessment.
B. Budgeting appropriately may come as a result, but is not the reason to perform a risk assessment.
C. Analyzing the effect on the business is part of the process, but the needs or acceptable effect or response must also be determined.
D. The risk assessment is used to identify and evaluate the impact of failure on critical business processes (and IT components supporting them) and to determine time frames, priorities, resources and interdependencies. It is part of the process to help determine the current state of risk and helps determine risk countermeasures in alignment with business objectives.

R2-61 Which of the following is **MOST** important when evaluating and assessing risk to an enterprise or business process?

A. Identification of controls that are currently in place to mitigate identified risk
B. Threat intelligence, including likelihood of identified threats
C. Historical risk assessment data
D. Control testing results

B is the correct answer.

Justification:
A. Identification of controls that are currently in place is an important part of the risk assessment process, but is not as important as threat intelligence.
B. One of the key requirements of effective risk assessment is its association and alignment with current intelligence that includes data on the likelihood of identified threats. The probability of risk being realized is one of the primary determinations of risk prioritization.
C. Historical risk assessment data are useful in understanding previously identified risk, but are not essential to the risk assessment process.
D. Control testing results are a component of risk assessment that helps support conclusions. Threat intelligence will often drive the testing of specific controls based on the identification of risk scenarios during the evaluation and assessment activity. These data are valuable to the risk assessment process, but are not as valuable as accurate threat intelligence.

R2-62 It is **MOST** important for a risk evaluation to:

A. take into account the potential size and likelihood of a loss.
B. consider inherent and control risk.
C. include a benchmark of similar companies in its scope.
D. assume an equal degree of protection for all assets.

A is the correct answer.

Justification:
A. Risk evaluation should take into account the potential size and likelihood of a loss.
B. Although inherent and control risk should be considered in the analysis, the impact of the risk (potential likelihood and impact of loss) should be the primary driver.
C. Risk evaluation can include comparisons with a group of companies of similar size.
D. Risk evaluation should not assume an equal degree of protection for all assets because assets may have different risk factors.

R2-63 When performing a risk assessment on the impact of losing a server, calculating the monetary value of the server should be based on the:

A. cost to obtain a replacement.
B. annual loss expectancy (ALE).
C. cost of the software stored.
D. original cost to acquire.

A is the correct answer.

Justification:

A. The value of the server should be based on its replacement cost; however, the financial impact to the enterprise may be much broader, based on the function that the server performs for the business and the value it brings to the enterprise.
B. The annual loss expectancy (ALE) for all risk related to the server does not represent the server's value.
C. The software can be restored from backup media.
D. The original cost may be significantly different from the current cost and, therefore, not as relevant.

R2-64 Which of the following factors should be included when assessing the impact of losing network connectivity for 18 to 24 hours?

A. The hourly billing rate charged by the carrier
B. Financial losses incurred by affected business units
C. The value of the data transmitted over the network
D. An aggregate compensation of all affected business users

B is the correct answer.

Justification:

A. The hourly billing rate charged by the carrier may be a factor that contributes to the overall financial impact; however, it is a very limited subset of the actual impact of losing network connectivity.
B. The impact of network unavailability is the cost it incurs to the enterprise.
C. The value of the data transmitted over the network is a subset of the financial losses incurred by affected business units.
D. An aggregate compensation of all affected business users is a subset of the financial losses incurred by affected business units.

R2-65 Which of the following is the **PRIMARY** factor when deciding between conducting a quantitative or qualitative risk assessment?

A. The corporate culture
B. The amount of time available
C. The availability of data
D. The cost involved with risk assessment

C is the correct answer.

Justification:
A. Management will make decisions based on the risk assessment provided. If management makes decisions based only on financial values, then a quantitative risk analysis is appropriate. If the decision will be based on non-numerical values regarding conceptual elements, then a qualitative analysis is appropriate.
B. The amount of time available may be a factor in deciding between a quantitative and qualitative analysis, but it is not the primary factor.
C. The availability of data is the primary factor in deciding between a quantitative and qualitative risk analysis.
D. The cost involved with a risk assessment may be a factor in deciding between a quantitative and qualitative analysis, but it is not the primary factor.

R2-66 Which of the following is the **BEST** method to analyze risk, incidents and related interdependencies to determine the impact on organizational goals?

A. Security information and event management (SIEM) solutions
B. A business impact analysis (BIA)
C. Enterprise risk management (ERM) steering committee meetings
D. Interviews with business leaders to develop a risk profile

B is the correct answer.

Justification:
A. Security information and event management (SIEM) solutions will primarily account for technical risk and typically do not evaluate the impact that business process objectives have on operational components.
B. A business impact analysis (BIA) should include the examination of risk, incidents and interdependencies as part of the activity to identify impact to business objectives.
C. Enterprise risk management (ERM) steering committees are useful for reviewing analyses that have been completed, but not for conducting analysis activities.
D. Interviews with business leaders will assist in identifying risk tolerance and key business objectives and activities, but will not analyze risk or incidents.

R2-67 A risk assessment process that uses likelihood and impact in calculating the level of risk is a:

A. qualitative process.
B. failure modes and effects analysis (FMEA).
C. fault tree analysis.
D. quantitative process.

D is the correct answer.

Justification:
A. A qualitative risk assessment process uses scenarios and ranking of risk levels in calculating the level of risk.
B. A failure modes and effects analysis (FMEA) determines the extended impact of an adverse event on other systems or operational areas.
C. A fault tree analysis risk assessment determines threats by considering all threat sources to a business process.
D. A quantitative risk assessment process uses likelihood and impact in calculating the monetary value of risk.

R2-68 Which of the following **BEST** describes the role of management in implementing a risk management strategy?

A. Ensure that the planning, budgeting and performance of information security components are appropriate.
B. Assess and incorporate the results of the risk management activity into the decision-making process.
C. Identify, evaluate and minimize risk to IT systems that support the mission of the organization.
D. Understand the risk management process so that appropriate training materials and programs can be developed.

B is the correct answer.

Justification:

A. Ensuring the planning, budgeting and performance of information security components is usually the responsibility of the chief information officer (CIO). Although the CIO is a member of senior management, this does not best describe the collective role of senior management in establishing and implementing the risk management strategy.
B. Assessing and incorporating the results of the risk management activity into the decision-making process best describes the role of senior management in establishing and implementing a risk management strategy.
C. Identifying, evaluating and minimizing risk to IT systems supporting the corporate mission is done by IT security managers or an IT security function, but this does not best describe the role of senior management in creating the risk management strategy.
D. Understanding the risk management process to develop appropriate training materials and programs is usually the role of corporate security trainers, not of senior management.

R2-69 Which of the following is the **GREATEST** challenge of performing a quantitative risk analysis?

A. Obtaining accurate figures on the impact of a realized threat
B. Obtaining accurate figures on the value of assets
C. Calculating the annual loss expectancy (ALE) of a specific threat
D. Obtaining accurate figures on the frequency of specific threats

D is the correct answer.

Justification:

A. The impact of a threat can be determined based on the type of threat that occurs.
B. The value of an asset should be fairly easy to ascertain.
C. Annual loss expectancy (ALE) will not be difficult to calculate if you know the correct frequency of the threat occurring.
D. It can be challenging to obtain an accurate figure on the frequency of a threat occurring.

R2-70 The **MOST** effective method to conduct a risk assessment on an internal system in an organization is to start by understanding the:

A. performance metrics and indicators.
B. policies and standards.
C. recent audit findings and recommendations.
D. system and its subsystems.

D is the correct answer.

Justification:
A. The person performing the risk assessment should already understand the performance metrics and indicators.
B. The person performing the risk assessment should already understand the policies and standards of the organization.
C. Recent audit findings and recommendations could be useful, but are not as important as understanding the system.
D. To conduct a proper risk assessment, the risk practitioner must understand the system, subsystems and how they work. This provides knowledge of how policies and standards are applied within the system and subsystems, an understanding of process-specific risk, existing interdependencies, and performance indicators.

R2-71 The board of directors wants to know the financial impact of specific, individual risk scenarios. What type of approach is **BEST** suited to fulfill this requirement?

A. Delphi method
B. Quantitative analysis
C. Qualitative analysis
D. Financial risk modeling

B is the correct answer.

Justification:
A. The Delphi method is a forecasting method based on expert opinions that are gathered in several, anonymous survey iterations.
B. A quantitative approach to risk evaluations would be the best approach because it is formula-based and puts a monetary amount on the potential loss resulting from a risk scenario.
C. Qualitative analysis does not quantify the risk and loss in numbers and therefore is not the best option.
D. Financial risk modeling is used to determine the aggregate risk in a financial portfolio. It is generally not used to provide the financial impact of individual risk scenarios.

R2-72 Which of the following objectives is the **PRIMARY** reason risk professionals conduct risk assessments?

A. To maintain the enterprise's risk register
B. To enable management to choose the right risk response
C. To provide assurance on the risk management process
D. To identify risk with the highest business impact

D is the correct answer.

Justification:
A. The maintenance of the risk register is part of the ongoing risk assessment process.
B. Management chooses the right risk response strategy based on risk analysis. A risk assessment itself is not sufficient to make educated risk response decisions.
C. Assurance on risk management is not the main reason risk assessment is performed by the risk professional.
D. A risk assessment is the process used to identify risk and develop risk scenarios to determine how specific threats may adversely affect the business.

R2-73 A new regulation for safeguarding information processed by a specific type of transaction has come to the attention of an IT manager. The manager should **FIRST**:

A. meet with stakeholders to decide how to comply.
B. analyze the key risk in the compliance process.
C. update the existing security/privacy policy.
D. assess whether existing controls meet the regulation.

D is the correct answer.

Justification:

A. Meeting with stakeholders is a subsequent action to understanding the impact and requirements and performing a gap assessment.
B. Analyzing the key risk in the compliance process is a subsequent action to understanding the impact and requirements and performing a gap assessment.
C. Updating the existing security/privacy policy is a subsequent action to the understanding the impact and requirements and performing a gap assessment.
D. The first step is to understand the impact and requirements of the new regulation, which includes assessing how the enterprise will comply with the regulation and to what extent the existing control structure supports the compliance process. The risk practitioner should then assess any existing gaps.

R2-74 How often should risk be evaluated?

A. Annually or when there is a significant change
B. Once a year for each business process and subprocess
C. Every three to six months for critical business processes
D. Only after significant changes occur

A is the correct answer.

Justification:

A. Risk is constantly changing. Evaluating risk annually or when there is a significant change offers the best alternative because it takes into consideration a reasonable time frame while allowing flexibility to address significant change.
B. Evaluating risk once a year is insufficient if important changes take place.
C. Evaluating risk every three to six months for critical processes may not be necessary or it may not address important changes in a timely manner.
D. Evaluating risk only after significant changes occur may not take into consideration the effect of time and less significant changes that may collectively affect the overall risk.

R2-75 Who is **MOST** likely responsible for data classification?

A. The data user
B. The data owner
C. The data custodian
D. The system administrator

B is the correct answer.

Justification:
A. The data user is granted access based on a justified business need and after approval from the data owner.
B. The data owner is responsible for classifying data according to the enterprise's data classification scheme. The classification scheme then defines who is eligible to access the data and what controls are required.
C. The data custodian is responsible for the safe custody, transport and storage of the data (and implementation of business rules).
D. System administrators are considered data custodians because they ensure the safe custody, transport and storage of the data (and implementation of business rules).

R2-76 Which of the following **BEST** describes the objective of a business impact analysis (BIA)?

A. The identification of threats, risk and vulnerabilities that can adversely affect the enterprise
B. The development of procedures for initial response and stabilization of situations during an emergency
C. The identification of time-sensitive critical business functions and interdependencies
D. The development of communication procedures in the case of a crisis impacting the business

C is the correct answer.

Justification:
A. The identification of threats, risk and vulnerabilities is the objective of risk identification and analysis.
B. The development of procedures for initial response and stabilization of situations during an emergency is a key output of preparedness and response planning.
C. Identification of time-sensitive critical business functions and interdependencies is a deliverable of the business impact analysis (BIA); this is reflected partially by metrics, such as recovery time objectives (RTOs) and recovery point objectives (RPOs).
D. Communication procedures are beneficial to every business process, including crisis management; however, they are not the main deliverable of the BIA and relate more closely to business continuity and disaster recovery planning.

R2-77 A **MAJOR** risk of using single sign-on (SSO) is that it:

A. uses complex technologies for password management.
B. may potentially bypass the enterprise firewall.
C. is prone to distributed denial-of-service (DDoS) attacks.
D. may be a potential single point of compromise.

D is the correct answer.

Justification:

A. Single sign-on (SSO) technologies have built-in routines for password management; once the application is configured, it is generally easy to manage by qualified personnel.
B. SSO deployments do not bypass an enterprise firewall; they generally involve a server that talks to each user's system and if the system is outside the network, specific firewall ports will be configured to allow inbound and outbound traffic.
C. SSO in itself is not prone to distributed denial-of-service (DDoS) attacks; it is a technology to enforce access to multiple applications and devices using a single username and password. If properly hardened, an SSO server will ensure that security attacks at the operating system (OS) or systemic level are controlled.
D. Because SSO uses a single username and password for access to multiple systems, loss or disclosure of this single username and password may cause compromise across a broad range of systems.

R2-78 Which of the following is **MOST** important in determining the risk mitigation strategy?

A. Review vulnerability assessment results.
B. Conduct a likelihood and impact ranking.
C. Perform a business impact analysis (BIA).
D. Align it with the security controls framework.

B is the correct answer.

Justification:

A. Results from a vulnerability assessment are used in determining the likelihood and impact, but the tolerance level must be predefined.
B. Ranking each risk based on the known (or perceived) impact and likelihood is critical in determining the risk mitigation strategy.
C. Business impact analysis (BIA) documents aid in mitigation and recovery strategy development because these documents explain processes, key deliverables and recovery time objectives (RTOs).
D. Understanding the enterprise's security controls framework assists in determining final treatments once the mitigation strategy is determined for a given risk.

R2-79 Which of the following risk management activities initially identifies critical business functions and key business risk?

A. Risk monitoring
B. Risk analysis
C. Risk assessment
D. Risk evaluation

C is the correct answer.

Justification:

A. Risk monitoring provides timely information on the actual status of the enterprise with regard to risk.
B. Risk analysis is a process by which frequency and magnitude of IT risk scenarios are estimated.
C. A risk assessment is a process used to identify and evaluate risk and its potential effects. It includes assessing the critical functions necessary for an enterprise to continue business operations, defining the controls in place to reduce enterprise exposure and evaluating the cost for such controls.
D. Risk evaluation is the process of comparing the estimated risk against given risk criteria to determine the significance of the risk.

R2-80 A lack of adequate controls represents:

A. an impact.
B. a risk indicator.
C. a vulnerability.
D. a threat.

C is the correct answer.

Justification:

A. Impact is the measure of the financial loss that a threat event may have.
B. A risk indicator is a metric capable of showing that the enterprise is subject to, or has a high probability of being subject to, a risk that exceeds the defined risk appetite.
C. The lack of adequate controls represents a vulnerability, exposing sensitive information and data to the risk of malicious damage, attack or unauthorized access by hackers. This could result in a loss of sensitive information, financial loss, legal penalties, etc.
D. A threat is a potential cause of an unwanted incident.

R2-81 A risk professional has been asked to determine which factors were responsible for a loss event. Which of the following methods should be used?

A. Key risk indicators (KRIs)
B. Cause-and-effect analysis
C. Business process modeling (BPM)
D. Business impact analysis (BIA)

B is the correct answer.

Justification:

A. Key risk indicators (KRIs) are a subset of risk indicators that are highly relevant and possess a high probability of predicting or indicating important risk. They are not used after a loss event occurs.

B. Cause-and-effect analysis is a predictive or diagnostic analytical tool used to explore the root causes or factors that contribute to positive or negative effects or outcomes. It can also be used to identify potential risk. A typical form is the fishbone diagram.

C. Business process modeling (BPM) is used to model business processes and is not used for root cause analysis.

D. Business impact analysis (BIA) is a process to determine the impact of losing the support of any resource and is not used for root cause analysis.

R2-82 Which of the following is the **BEST** method to ensure the overall effectiveness of a risk management program?

A. Assignment of risk within the enterprise
B. Comparison of the program results with industry standards
C. Participation by applicable members of the enterprise
D. User assessment of changes in risk

C is the correct answer.

Justification:

A. Assignment of risk within the enterprise is important to ensure that risk owners are clearly defined and aware of their responsibilities.

B. Comparison of the program results with industry standards may result in valuable feedback—similar to benchmarking—but is not as important as stakeholder participation.

C. Effective risk management requires participation, support and acceptance by all applicable members of the enterprise, beginning with the executive levels. Personnel must understand their responsibilities and be trained on how to fulfill their roles.

D. User assessment of changes is a subjective method of assessing risk and not part of a mature risk management program.

R2-83 The **PRIMARY** result of a risk management process is:

A. a defined business plan.
B. input for risk-aware decisions.
C. data classification.
D. minimized residual risk.

B is the correct answer.

Justification:
A. Risk management deliverables are not the primary input into the business plan.
B. Risk management identifies and prioritizes risk and relates the aggregated risk to the enterprise's risk appetite and risk tolerance levels to enable risk-aware decision making.
C. Establishing classification levels is one of the outputs of the outcome of risk assessment, but is not the primary result.
D. Residual risk is reduced after taking the cost of the risk response and the related benefit into consideration; risk minimization itself is not a primary result of risk management because it may not optimize overall business results.

R2-84 Investments in risk management technologies should be based on:

A. audit recommendations.
B. vulnerability assessments.
C. business climate.
D. value analysis.

D is the correct answer.

Justification:
A. Basing decisions on audit recommendations is reactive in nature and may not comprehensively address the key business needs.
B. Vulnerability assessments are useful, but they do not determine whether the cost is justified.
C. Demonstrated value takes precedence over the current business climate because the climate is ever changing.
D. Investments in risk management technologies should be based on a value analysis and sound business case.

R2-85 Senior management has defined the enterprise risk appetite as moderate. A business critical application has been determined to pose a high risk. What is the **BEST** next course of action?

A. Remove the high-risk application and replace it with another system.
B. Request that senior management increase the level of risk they are willing to accept.
C. Determine whether new controls to be implemented on the system will mitigate the high risk.
D. Restrict access to the application to trusted users.

C is the correct answer.

Justification:
A. Removing the application may lead to unacceptable downtime for a critical business function.
B. Requesting that senior management accept the high risk should only be done after functioning and compensating controls are factored in.
C. The risk practitioner should determine whether new controls to be implemented on the system may lower the risk from high to moderate or low before taking any further action.
D. Restricting the access to trusted users may not mitigate the high risk that the application poses.

R2-86 Which of the following is an example of postincident response activity?

A. Performing a cost-benefit analysis of corrective controls deployed for the incident
B. Reassessing the risk to make necessary amendments to procedures and guidelines
C. Removing the relevant security policies that resulted in increased incidents
D. Inviting the internal audit department to review the corrective controls

B is the correct answer.

Justification:

A. Performing a cost-benefit analysis of corrective controls deployed for the incident does not make sense because the corrective controls have already been deployed and the cost-benefit analysis is performed to select the best control.
B. In incident response, a response occurs in reaction to an incident. Risk analysis should be performed to learn from the incident and find long-term corrective actions, such as policy amendments, to reduce the probability and frequency of the incident reoccurring.
C. By removing security policies, the incident cannot be corrected and in fact may further escalate the risk.
D. Corrective controls are reviewed by the risk manager. Review from internal audit as a routine process can create conflict.

R2-87 Which of the following approaches is the **BEST** approach to exception management?

A. Escalation processes are defined.
B. Process deviations are not allowed.
C. Decisions are based on business impact.
D. Senior management judgment is required.

A is the correct answer.

Justification:

A. Ideally, each policy should include an escalation process, where exceptions are managed at the appropriate level of authority.
B. It is not feasible to rigidly mandate that process deviations are not allowed. Ideally, each policy should include an escalation process, where exceptions are escalated to the appropriate level of authority.
C. While decisions regarding exceptions should be based on impact, frequency of exceptions should also be considered. A simplistic example may be an outsourced hosting environment where the customer is charged for each password reset and password resets occur at an exceedingly high rate.
D. Senior management may not be in the best position to approve each exception. Ideally, each policy and procedure should include an escalation process, where exceptions are escalated to the appropriate level of authority.

R2-88 Which of the following factors should be analyzed to help management select an appropriate risk response?

A. The impact on the control environment
B. The likelihood of a given threat
C. The costs and benefits of the controls
D. The severity of the vulnerabilities

C is the correct answer.

Justification:
A. The impact of a risk response on the control environment needs to be balanced against the cost of the risk response.
B. The likelihood of a threat will not determine the selection of a risk response.
C. Analysis of costs and benefits for the controls would help to select an appropriate risk response. If the cost outweighs the benefits, the risk response selected is not optimal for the enterprise.
D. The severity of a vulnerability will not determine the selection of a risk response.

R2-89 The **GREATEST** risk to token administration is:

A. the ability to easily tamper with or steal a token.
B. the loss of network connectivity to the authentication system.
C. the inability to secure unassigned tokens.
D. the ability to generate temporary codes to log in without a token.

B is the correct answer.

Justification:
A. Tokens are built to be tamper proof, and it is very difficult to alter or duplicate a token.
B. A centrally controlled token-based system relies on secure and reliable network communications in order to authenticate the user at the time of login. A network or communications failure can lead to a denial of service or unauthorized access.
C. Unassigned tokens (physical and logical) must be properly secured to prevent unauthorized use. Tokens can provide access, but they require a valid password to grant access to the system.
D. The use of temporary codes is necessary when users forget their tokens; however, this work-around should be used only when the user is known to be authorized and valid.

R2-90 A process by which someone logs onto a web site, then receives a token via a short message service (SMS) message, is an example of what control type?

A. Deterrent
B. Directive
C. Compensating
D. Preventive

D is the correct answer.

Justification:
A. A deterrent control will discourage improper behavior, but not prevent it.
B. A directive control guides behavior, but will not prevent unauthorized access.
C. A compensating control addresses a weakness in other controls, but the use of a token-based system will provide adequate control.
D. The use of a token with a short message service (SMS) message will prevent unauthorized access to the system through two-factor authentication.

R2-91 Which of the following actions will an incident response plan activation **MOST** likely involve?

A. Enabling logging to track what resources have been accessed
B. Shutting down a server to patch defects in the operating system
C. Implementing virus scanning tools to scan attachments in incoming email
D. Assisting in the migration to an alphanumeric password authorization policy

B is the correct answer.

Justification:

A. Enabling logging is not a function of the incident response plan, but can provide information needed if it has been enabled prior to the incident.
B. An incident response plan includes actions to react to a threat, loss or vulnerability event. The shutting down of servers to patch defects is a corrective action against identified events. Once installed, the upgraded version of the operating system might be able to prevent further risk from materializing.
C. Use of a virus scanner is a preventive and a detective action rather than correcting what has occurred. Use of a virus scanner in the response to an incident will mean scanning email that has already been received.
D. Generally, an alphanumeric password authorization policy is a preventive rather than a corrective control.

R2-92 Which of the following is of **MOST** concern in a review of a virtual private network (VPN) implementation? Computers on the network are located:

A. at the enterprise's remote offices.
B. on the enterprise's internal network.
C. at the backup site.
D. in employees' homes.

D is the correct answer.

Justification:

A. Computers on the network that are at the enterprise's remote offices, perhaps with different IS and security employees who have different ideas about security, may be more risky, but are less risky than an employee's home computer.
B. There should be security policies in place on an enterprise's internal network to detect and halt an outside attack that uses an internal machine as a staging platform.
C. Computers at the backup site are subject to the corporate security policy and, therefore, are not high-risk computers.
D. In a virtual private network (VPN) implementation, there is a risk of allowing high-risk computers onto the enterprise's network. All machines that are allowed onto the virtual network should be subject to the same security policy. Home computers are least often subject to the corporate security policies and, therefore, are high-risk computers. Once a computer is hacked and "owned," any network that trusts that computer is at risk. Implementation and adherence to the corporate security policy is easier when all computers on the network are on the enterprise's campus.

R2-93 Which type of risk assessment methods involves conducting interviews and using anonymous questionnaires by subject matter experts?

A. Quantitative
B. Probabilistic
C. Monte Carlo
D. Qualitative

D is the correct answer.

Justification:
A. Quantitative risk assessments utilize a mathematical calculation based on security metrics on the asset (system or application).
B. Probabilistic risk assessments use a mathematical model to construct the qualitative risk assessment approach while using the quantitative risk assessment techniques and principles.
C. Monte Carlo simulation combines both qualitative and quantitative assessment approaches and is based on a normal deterministic simulation model.
D. Qualitative risk assessment methods include using interviewing and the Delphi method, which is the method described in the question.

R2-94 Which of the following is the **MOST** prevalent risk in the development of end-user computing (EUC) applications?

A. Increased development and maintenance costs
B. Increased application development time
C. Impaired decision making due to diminished responsiveness to requests for information
D. Applications not subjected to testing and IT general controls

D is the correct answer.

Justification:
A. End-user computing (EUC) applications typically result in reduced application development and maintenance costs.
B. EUC applications typically result in a reduced development cycle time.
C. EUC applications normally increase flexibility and responsiveness to management's information requests.
D. End-user-developed applications may not be subject to an independent outside review by systems analysts and, frequently, are not created in the context of a formal development methodology. These applications may lack appropriate standards, controls, quality assurance procedures and documentation. A risk associated with end-user applications can include backup and recovery procedures that are not applied because operations may not be aware of the application.

R2-95 The annual expected loss of an asset—the annual loss expectancy (ALE)—is calculated as the:

A. exposure factor (EF) multiplied by the annualized rate of occurrence (ARO).
B. single loss expectancy (SLE) multiplied by the exposure factor (EF).
C. single loss expectancy (SLE) multiplied by the annualized rate of occurrence (ARO).
D. asset value (AV) multiplied by the single loss expectancy (SLE).

C is the correct answer.

Justification:
A. This is not the correct formula to calculate annual loss expectancy (ALE). ALE is calculated by multiplying the single loss expectancy (SLE) by the annualized rate of occurrence (ARO) or the amount of times that the enterprise expects the loss to occur.
B. This is not the correct formula to calculate ALE. ALE is calculated by multiplying the SLE by the ARO or the amount of times that the enterprise expects the loss to occur.
C. ALE is calculated by is calculated by multiplying the SLE by the ARO or the amount of times that the enterprise expects the loss to occur.
D. This is not the correct formula to calculate ALE. ALE is calculated by multiplying the SLE by the ARO or the amount of times that the enterprise expects the loss to occur.

R2-96 Which of the following is **MOST** useful when computing annual loss exposure?

A. The cost of existing controls
B. The number of vulnerabilities
C. The net present value (NPV) of the asset
D. The business value of the asset

D is the correct answer.

Justification:
A. The cost of existing controls is not taken into consideration for calculation of the annual loss exposure.
B. The number of vulnerabilities does not help determine the annual loss exposure.
C. Net present value (NPV) is based on asset depreciation value and is a difficult basis for annual loss exposure because it may not reflect the true risk associated with the asset.
D. Annual loss exposure is a function of value of the information asset involved and the risk impact if the potential risk materializes. This needs to be identified primarily to determine the exposure factor of the options provided in this question.

R2-97 Which of the following techniques **BEST** helps determine whether there have been unauthorized program changes since the last authorized program update?

A. A test data run
B. An automated code comparison
C. A code review
D. A review of code migration procedures

B is the correct answer.

Justification:

A. Test data runs help verify the processing of preselected transactions, but provide no evidence about unexercised portions of a program.
B. An automated code comparison is the process of comparing two versions of the same program to determine whether the two correspond. It is an efficient technique because it is an automated procedure.
C. A code review is the process of reading program source code listings to determine whether the code contains potential errors or inefficient statements. A code review can be used as a means of code comparison, but it is not efficient.
D. A review of code migration procedures would not detect program changes.

R2-98 What indicates that an enterprise's risk practices need to be reviewed?

A. The IT department has its own methodology of risk management.
B. Manufacturing assigns its own internal risk management roles.
C. The finance department finds exceptions during its yearly risk review.
D. Sales department risk management procedures were last reviewed 11 months ago.

A is the correct answer.

Justification:

A. An enterprise needs to use the same methodology for risk management across the enterprise in order to properly identify and report on risk.
B. Assigning internal risk management roles to staff is what each department in the enterprise should do.
C. It is common to find exceptions during a review that need to be addressed. This is a normal and expected result of a yearly review.
D. Normally, a yearly review of risk management procedures is sufficient for enterprises to keep them up to date.

Page intentionally left blank

DOMAIN 3—RISK RESPONSE AND MITIGATION (23%)

R3-1 Because of its importance to the business, an enterprise wants to quickly implement a technical solution that deviates from the company's policies. The risk practitioner should:

A. recommend against implementation because it violates the company's policies.
B. recommend revision of the current policy.
C. conduct a risk assessment and allow or disallow based on the outcome.
D. recommend a risk assessment and subsequent implementation only if residual risk is accepted.

D is the correct answer.

Justification:
A. Every business decision is driven by cost and benefit considerations. A risk practitioner's contribution to the process is most likely a risk assessment, identifying both the risk and opportunities related to the proposed solution.
B. A recommendation to revise the current policy should not be triggered by a single request.
C. While a risk practitioner may conduct a risk assessment to enable a risk-aware business decision, it is management who will make the final decision.
D. A risk assessment should be conducted to clarify the risk whenever the company's policies cannot be followed. The solutions should only be implemented if the related risk is formally accepted by the business.

R3-2 When proposing the implementation of a specific risk mitigation activity, a risk practitioner **PRIMARILY** utilizes a:

A. technical evaluation report.
B. business case.
C. vulnerability assessment report.
D. budgetary requirements.

B is the correct answer.

Justification:
A. A technical evaluation report is supplemental to a business case.
B. A manager needs to base the proposed risk response on a risk evaluation, the business need (new product, changes in process, compliance need, etc.) and the requirements of the enterprise (new technology, cost, etc.). The manager must look at the costs of the various controls and compare them against the benefit that the organization will receive from the risk response. The manager needs to have knowledge of business case development to illustrate the costs and benefits of the risk response.
C. A vulnerability assessment report is supplemental to a business case.
D. Budgetary requirements are an input into a business case.

R3-3 Risk management programs are designed to reduce risk to:

A. the point at which the benefit exceeds the expense.
B. a level that is too small to be measurable.
C. a rate of return that equals the current cost of capital.
D. a level that the enterprise is willing to accept.

D is the correct answer.

Justification:
A. Depending on the risk preference of an enterprise, it may or may not choose to pursue risk mitigation to the point at which the benefit equals or exceeds the expense.
B. Reducing risk to a level too small to measure is not practical and is often cost-prohibitive.
C. Tying risk to a specific rate of return ignores the qualitative aspects of risk that must also be considered.
D. Risk should be reduced to a level that an organization is willing to accept.

R3-4 Whether a risk has been reduced to an acceptable level should be determined by:

A. IS requirements.
B. information security requirements.
C. international standards.
D. organizational requirements.

D is the correct answer.

Justification:
A. IS requirements should not make the ultimate determination.
B. Information security requirements should not make the ultimate determination.
C. Because each enterprise is unique, international standards of best practice do not represent the best solution.
D. Organizational requirements should determine when a risk has been reduced to an acceptable level.

R3-5 Which of the following is the **MOST** effective way to treat a risk such as a natural disaster that has a low probability and a high impact level?

A. Eliminate the risk.
B. Accept the risk.
C. Transfer the risk.
D. Implement countermeasures.

C is the correct answer.

Justification:
A. Eliminating the risk may not be possible.
B. Accepting the risk leaves the enterprise vulnerable to a catastrophic disaster that could cripple or ruin the organization.
C. Typically, when the probability of an incident is low, but the impact is high, risk is transferred to insurance companies. Examples include hurricanes, tornados and earthquakes. While an enterprise cannot technically transfer risk, transferring risk describes a risk response in which an enterprise indemnifies against the impact of the realized risk.
D. Implementing countermeasures may not be the most cost-effective approach to security management. It would be more cost-effective to pay recurring insurance costs than to be affected by a disaster from which the enterprise cannot financially recover.

R3-6 A risk response report includes recommendations for:

A. acceptance.
B. assessment.
C. evaluation.
D. quantification.

A is the correct answer.

Justification:
A. Acceptance of a risk is an alternative to be considered in the risk response process.
B. The risk assessment process is completed prior to determining appropriate risk responses.
C. Risk evaluation is part of the risk assessment process that is completed prior to determining appropriate risk responses.
D. Risk quantification is achieved during risk analysis; it is an input into the risk response process and occurs prior to determining appropriate risk responses.

R3-7 Which of the following is minimized when acceptable risk is achieved?

A. Transferred risk
B. Control risk
C. Residual risk
D. Inherent risk

C is the correct answer.

Justification:
A. Transferred risk is risk that has been shared with a third party, such as an insurance provider; it may not necessarily be equal to the minimal amount of residual risk.
B. Control risk is the risk that controls may not meet the control objective.
C. After putting into place an effective risk management program, the remaining risk is called residual risk. Acceptable risk is achieved when residual risk is minimized.
D. Inherent risk is a risk that is part of an activity; it cannot be minimized, only avoided by not engaging in the activity itself.

R3-8 A global financial institution has decided not to take any further action on a denial-of-service (DoS) vulnerability found by the risk assessment team. The **MOST** likely reason for making this decision is that:

A. the needed countermeasure is too complicated to deploy.
B. there are sufficient safeguards in place to prevent this risk from happening.
C. the likelihood of the risk occurring is unknown.
D. the cost of countermeasure outweighs the value of the asset and potential loss.

D is the correct answer.

Justification:
A. While countermeasures can be too complicated to deploy, this is not the most compelling reason.
B. Any safeguards placed to prevent the risk need to match the risk impact.
C. It is likely that a global financial institution may be exposed to such denial-of-service (DoS) attacks, but the frequency cannot be predicted.
D. An enterprise may decide to accept a specific risk because the protection would cost more than the potential loss.

R3-9 Which of the following is **MOST** relevant to include in a cost-benefit analysis of a two-factor authentication system?

A. The approved budget of the project
B. The frequency of incidents
C. The annual loss expectancy (ALE) of incidents
D. The total cost of ownership (TCO)

D is the correct answer.

Justification:

A. The approved budget of the project may have no bearing on what the project may actually cost.
B. The frequency and annual loss expectancy (ALE) of incidents can help measure the benefit, but have more of an indirect relationship because not all incidents may be mitigated by implementing a two-factor authentication system.
C. The frequency and ALE of incidents can help measure the benefit, but have more of an indirect relationship because not all incidents may be mitigated by implementing a two-factor authentication system.
D. Total cost of ownership (TCO) is the most relevant piece of information to be included in the cost-benefit analysis because it establishes a cost baseline that must be considered for the full life cycle of the control.

R3-10 In the risk management process, a cost-benefit analysis is **MAINLY** performed:

A. as part of an initial risk assessment.
B. as part of risk response planning.
C. during an information asset valuation.
D. when insurance is calculated for risk transfer.

B is the correct answer.

Justification:

A. A cost-benefit analysis is not only performed once, but every time controls need to be selected to address new or changing risk.
B. In risk response, a range of controls will be identified that can mitigate the risk; however, a cost-benefit analysis in this process will help identify the right controls that will address the risk at acceptable levels within the budget.
C. In information asset valuation, business owners determine the value based on business importance and there is no cost-benefit involved.
D. Calculating insurance for the purpose of transferring risk is not the stage where a cost-benefit analysis is performed.

R3-11 During a risk management exercise, an analysis was conducted on the identified risk and mitigations were identified. Which choice **BEST** reflects residual risk?

A. Risk left after the implementation of new or enhanced controls
B. Risk mitigated as a result of the implementation of new or enhanced controls
C. Risk identified prior to implementation of new or enhanced controls
D. Risk classified as high after the implementation of new or enhanced controls

A is the correct answer.

Justification:
A. The classic definition of residual risk is any risk left after appropriate controls have been implemented to mitigate the target risk.
B. Residual risk is the risk that remains and is not mitigated.
C. Risk is not identified prior, but after, the implementation of controls.
D. Residual risk can be rated at any level, not just high risk.

R3-12 Which of the following choices will **BEST** protect the enterprise from financial risk?

A. Insuring against the risk
B. Updating the IT risk registry
C. Improving staff training in the risk area
D. Outsourcing the process to a third party

A is the correct answer.

Justification:
A. An insurance policy can compensate the enterprise up to 100 percent.
B. Updating the risk registry (with lower values for impact and probability) will not change the risk, only management's perception of it.
C. Staff capacity to detect or mitigate the risk may potentially reduce the financial impact, but never cover it fully the way insurance can.
D. Outsourcing the process containing the risk does not necessarily remove or change the risk.

R3-13 After the completion of a risk assessment, it is determined that the cost to mitigate the risk is much greater than the benefit to be derived. A risk practitioner should recommend to business management that the risk be:

A. treated.
B. terminated.
C. accepted.
D. transferred.

C is the correct answer.

Justification:
A. Treating the risk in the described scenario incurs a cost that is greater than the benefit to be derived; this is not the best option.
B. Risk termination is not a risk management term; while risk can be avoided, it can generally not be terminated.
C. When the cost of control is more than the cost of the potential impact, the risk should be accepted.
D. Transferring risk is of limited benefit if the cost of the risk response is more than the cost of the potential likelihood and impact of the risk itself.

R3-14 A **PRIMARY** reason for initiating a policy exception process is when:

A. the risk is justified by the benefit.
B. policy compliance is difficult to enforce.
C. operations are too busy to comply.
D. users may initially be inconvenienced.

A is the correct answer.

Justification:

A. **Exceptions to policy are warranted in circumstances in which the benefits outweigh the costs of policy compliance; however, the enterprise needs to asses both the tangible and intangible risk and assess those against the existing risk.**
B. Difficult enforcement is not justification for policy exceptions.
C. Busy operations are not justification for policy exceptions.
D. User inconvenience is not a reason to automatically grant exception to a policy.

R3-15 A risk practitioner receives a message late at night that critical IT equipment will be delivered several days late due to flooding. Fortunately, a reciprocal agreement exists with another company for a replacement until the equipment arrives. This is an example of risk:

A. transfer.
B. avoidance.
C. acceptance.
D. mitigation.

D is the correct answer.

Justification:

A. Risk transfer is not the correct answer because the described risk is not transferred using insurance or another risk transfer strategy.
B. Arranging for a standby is a risk mitigation strategy, not a risk avoidance strategy.
C. The risk is not accepted; if it were accepted, the enterprise would, for example, continue operating without the expected IT equipment until it was delivered.
D. **Risk mitigation attempts to reduce the impact when a risk event occurs. Making plans such as a reciprocal arrangement with another company reduces the consequence of the risk event.**

R3-16 Which of the following would **BEST** help an enterprise select an appropriate risk response?

A. The degree of change in the risk environment
B. An analysis of risk that can be transferred were it not eliminated
C. The likelihood and impact of various risk scenarios
D. An analysis of control costs and benefits

D is the correct answer.

Justification:

A. The degree of change in the risk environment will not provide information of actual controls and benefits to make the decision.
B. Risk can never be eliminated and even analysis of what risk can be transferred will be inadequate for a complete risk response.
C. Likelihood and impact help establish the amount or level of risk.
D. **An analysis of costs and benefits for controls helps an enterprise understand if it can mitigate the risk to an acceptable level.**

R3-17 Which of the following leads to the **BEST** optimal return on security investment?

A. Deploying maximum security protection across all of the information assets
B. Focusing on the most important information assets and then determining their protection
C. Deploying minimum protection across all the information assets
D. Investing only after a major security incident is reported to justify investment

B is the correct answer.

Justification:
A. Deploying maximum controls across all information assets will overprotect some of the less critical information assets; therefore, investment will not be optimized.
B. To optimize return on security investment, the primary focus should be identifying the important information assets and protecting them appropriately to optimize investment (i.e., important information assets get more protection than less important or critical assets).
C. Deploying minimum protection across all the information assets will compromise the security of the more critical information assets; therefore, investment will not be optimized.
D. Investing only after a major security event is a reactive approach that may severely compromise business operations—in some cases, to the extent where the business does not survive.

R3-18 As part of fire drill testing, designated doors swing open, as planned, to allow employees to leave the building faster. An observer notices that this practice allows unauthorized personnel to enter the premises unnoticed. The **BEST** way to alter the process is to:

A. stop the designated doors from opening automatically in case of a fire.
B. include the local police force to guard the doors in case of fire.
C. instruct the facilities department to guard the doors and have staff show their badge when exiting the building.
D. assign designated personnel to guard the doors once the alarm sounds.

D is the correct answer.

Justification:
A. Stopping the doors from opening in case of a fire does not effectively support the primary objective of a fire drill, which is to protect human life in case a fire occurs.
B. This choice is not useful because the police have better things to do.
C. Having the facilities department guard the exit doors and monitor staff as they leave the facility does not address the risk of having unauthorized personnel entering the building.
D. Unless there are designated personnel monitoring each door from the time the alarm sounds, there is no way to prevent unauthorized individuals from entering the building while employees are exiting.

R3-19 During a quarterly interdepartmental risk assessment, the IT operations center indicates a heavy increase of malware attacks. Which of the following recommendations to the business is **MOST** appropriate?

A. Contract with a new anti-malware software vendor because the current solution seems ineffective.
B. Close down the Internet connection to prevent employees from visiting infected web sites.
C. Make the number of malware attacks part of each employee's performance metrics.
D. Increase employee awareness training, including end-user roles and responsibilities.

D is the correct answer.

Justification:

A. Anti-malware software is always a step behind the malware that exists in the marketplace. This is particularly true for zero-day exploits; because the IT operation center is aware of the attack, the anti-malware in place seems to be effective.
B. Closing down the Internet connections may impair valid business processes and does not provide protection from the variety of channels that malware uses for attack.
C. Making employees responsible for the number of malware attacks that the enterprise is exposed to is an example of incentive misalignment because it punishes employees for something for which they are not responsible.
D. Employee awareness training will help the enterprise avoid, and more quickly detect, malware attacks, particularly when staff understand the typical symptoms and are knowledgeable about the incident reporting process.

R3-20 In a situation where the cost of anti-malware exceeds the loss expectancy of malware threats, what is the **MOST** viable risk response?

A. Risk elimination
B. Risk acceptance
C. Risk transfer
D. Risk mitigation

B is the correct answer.

Justification:

A. Risk elimination is not a risk response because it is not possible to reduce risk to zero.
B. When the cost of a risk response (i.e., the implementation of anti-malware) exceeds the loss expectancy, the most viable risk response is risk acceptance.
C. Transferring risk to a third party is most viable in situations where the potential likelihood is low and the potential impact is high. Transfer of risk—like any risk response—must be based on a cost-benefit analysis. If the cost of the risk exceeds the cost of the expected loss, the most viable risk response is to accept the risk.
D. Risk mitigation is a method to reduce the likelihood and/or impact of risk to an acceptable level. Risk mitigation—like any risk response—must be based on a cost-benefit analysis. If the cost of the risk exceeds the cost of the expected loss, the most viable risk response is to accept the risk.

R3-21 Which of the following is a behavior of risk avoidance?

A. Take no action against the risk.
B. Outsource the related process.
C. Insure against a specific event.
D. Exit the process that gives rise to risk.

D is the correct answer.

Justification:
A. Taking no action is an example of risk acceptance where no action is taken relative to a particular risk, and loss is accepted when/if it occurs. This is different from being ignorant of risk; accepting risk assumes that the risk is known, i.e., an informed decision has been made by management to accept it as such.
B. Outsourcing a process is an example of risk transfer/sharing. It reduces risk frequency or impact by transferring or otherwise sharing a portion of the risk. In both a physical and legal sense this risk transfer does not relieve an enterprise of a risk, but can involve the skills of another party in managing the risk and thus reduce the financial consequence if an adverse event occurs.
C. Insuring against a specific event is an example of risk transfer/sharing. It reduces risk frequency or impact by transferring or otherwise sharing a portion of the risk. In both a physical and legal sense risk transfer does not relieve an enterprise of a risk, but can involve the skills of another party in managing the risk and thus reduce the financial consequence if an adverse event occurs.
D. Avoidance means exiting the activities or conditions that give rise to risk. Risk avoidance applies when no other risk response is adequate. Some IT-related examples of risk avoidance may include relocating a data center away from a region with significant natural hazards or declining to engage in a very large project when the business case shows a notable risk of failure.

R3-22 Which of the following is **MOST** important for determining what security measures to put in place for a critical information system?

A. The number of threats to the system
B. The level of acceptable risk to the enterprise
C. The number of vulnerabilities in the system
D. The existing security budget

B is the correct answer.

Justification:
A. Determining the number of threats to the system is important; however, it alone will not determine the security measures to put in place.
B. Determining the level of acceptable risk will allow the enterprise to determine the security measures to put in place.
C. Determining the number of vulnerabilities in the system is important; however, it alone will not determine the security measures to put in place.
D. Determining how much of the budget is available should not determine the security measures to put in place.

R3-23 A chief information security officer (CISO) has recommended several controls such as anti-malware to protect the enterprise's information systems. Which approach to handling risk is the CISO recommending?

A. Risk transference
B. Risk mitigation
C. Risk acceptance
D. Risk avoidance

B is the correct answer.

Justification:

A. Risk transfer involves transferring the risk to another entity such as an insurance company.
B. By implementing controls the company is trying to decrease risk to an acceptable level, thereby mitigating risk.
C. Risk acceptance involves making an educated decision to accept the risk in the system and taking no action.
D. Risk avoidance involves stopping any activity causing the risk.

R3-24 Obtaining senior management commitment and support for information security investments can **BEST** be accomplished by a business case that:

A. explains the technical risk to the enterprise.
B. includes industry best practices as they relate to information security.
C. details successful attacks against a competitor.
D. ties security risk to organizational business objectives.

D is the correct answer.

Justification:

A. Senior management will not be as interested in technical risk if they are not tied to the impact on business environment and objectives.
B. Industry best practices are important to senior management but, again, senior management will give them the right level of importance when they are presented in terms of key business objectives.
C. Senior management will not be as interested in examples of successful attacks against a competitor if they are not tied to the impact on business environment and objectives.
D. Senior management seeks to understand the business justification for investing in security. This can best be accomplished by tying security to key business objectives.

R3-25 Acceptable risk for an enterprise is achieved when:

A. transferred risk is minimized.
B. control risk is minimized.
C. inherent risk is minimized.
D. residual risk is within tolerance levels.

D is the correct answer.

Justification:

A. Risk transfer is the process of assigning risk to another organization, usually through the purchase of an insurance policy or outsourcing the service. In both a physical and legal sense this risk transfer does not relieve an enterprise of a risk, but can involve the skills of another party in managing the risk and thus reduce the financial consequence if an adverse event occurs.
B. Control risk is the risk that a material error exists that would not be prevented or detected on a timely basis by the system of internal controls.
C. Inherent risk is the risk level or exposure without taking into account the actions that management has taken or might take (e.g., implementing controls). Inherent risk cannot be minimized.
D. Residual risk is the risk that remains after all controls have been applied; therefore, acceptable risk is achieved when residual risk is aligned with the enterprise risk appetite.

R3-26 A procurement employee notices that new printer models offered by the vendor keep a copy of all printed documents on a built-in hard disk. Considering the risk of unintentionally disclosing confidential data, the employee should:

A. proceed with the order and configure printers to automatically wipe all the data on disks after each print job.
B. notify the security manager to conduct a risk assessment for the new equipment.
C. seek another vendor that offers printers without built-in hard disk drives.
D. procure printers with built-in hard disks and notify staff to wipe hard disks when decommissioning the printer.

B is the correct answer.

Justification:
A. Wiping hard disks after each job is not appropriate without a prior risk assessment because the data may be useful for forensic investigation and may impact performance of the printer.
B. Risk assessment is the most appropriate answer because it will result in risk mitigation techniques that are appropriate for organizational risk context and appetite.
C. Focusing solely on the risk and ignoring the opportunity is not a correct approach. A risk associated with nonvolatile storage is not a sufficient reason for changing vendors because this is a general trend with the printers that brings business benefits in addition to risk that needs to be addressed.
D. Notifying staff is not a sufficient control and does not mitigate the risk from printers being serviced by an external party.

R3-27 Which of the following situations is **BEST** addressed by transferring risk?

A. An antiquated fire suppression system in the computer room
B. The threat of disgruntled employee sabotage
C. The possibility of the loss of a universal serial bus (USB) removable media drive
D. A building located in a 100-year flood plain

D is the correct answer.

Justification:
A. Although an enterprise may hold insurance policies for physical assets such as computer equipment, the most appropriate risk remediation strategy is to update the fire suppression system.
B. This risk is not readily transferrable. Full risk response planning should be performed for all risk that could happen at any time during routine business activities.
C. This risk is not readily transferrable. Removable media policies and procedures should proactively be in place to mitigate the risk of lost or stolen media.
D. Purchasing an insurance policy transfers the risk of a flood. Risk transfer is the process of assigning risk to another entity, usually through the purchase of an insurance policy or outsourcing the service.

SCENARIO 1

A scenario is a mini-case study that describes a situation or an organization and requires candidates to answer one or more questions based on the information provided. A scenario can focus on a specific domain or on several domains. The CRISC exam will include scenarios.

QUESTIONS R3-28 THROUGH R3-29 REFER TO THE FOLLOWING INFORMATION:

The chief information officer (CIO) of an enterprise has just received this year's IT security audit report. The report shows numerous open vulnerability findings on both business-critical and nonbusiness-critical information systems. The CIO briefed the chief executive officer (CEO) and board of directors on the findings and expressed his concern on the impact to the enterprise. He was informed that there are not enough funds to mitigate all of the findings from the report.

R3-28 The CIO should respond to the findings identified in the IT security audit report by mitigating:

A. the most critical findings on both the business-critical and nonbusiness-critical systems.
B. all vulnerabilities on business-critical information systems first.
C. the findings that are the least expensive to mitigate first to save funds.
D. the findings that are the most expensive to mitigate first and leave all others until more funds become available.

B is the correct answer.

Justification:
A. Mitigating the critical findings on the nonbusiness-critical systems is a waste of limited funds.
B. Mitigating vulnerabilities on business-critical information systems should be completed first to ensure that the business can continue to operate.
C. The expense of the mitigations should be a secondary factor to the value of the information systems.
D. The expense of the mitigations should be a secondary factor to the value of the information systems.

SEE INFORMATION PRECEDING QUESTION R3-29

R3-29 Assuming that the CIO is unable to address all of the findings, how should the CIO deal with any findings that remain after available funds have been spent?

A. Create a plan of actions and milestones for open vulnerabilities.
B. Shut down the information systems with the open vulnerabilities.
C. Reject the risk on the open vulnerabilities.
D. Implement compensating controls on the systems with open vulnerabilities.

A is the correct answer.

Justification:
A. Creating a plan of actions and milestones ensures that there is a plan to mitigate the remaining vulnerabilities over time. It will also identify the order in which the vulnerabilities should be mitigated with target dates for mitigation.
B. Shutting down the system is not the correct vulnerability mitigation strategy. The vulnerability may be on a mission-critical system.
C. Rejecting the risk is not a risk mitigation strategy.
D. Compensating controls will already be placed on the information systems. Additional compensating controls will require funds which have already been depleted.

R3-30 Which of the following **MOST** likely indicates that a customer data warehouse should remain in-house rather than be outsourced to an offshore operation?

A. The telecommunications costs may be much higher in the first year.
B. Privacy laws may prevent a cross-border flow of information.
C. Time zone differences may impede communications between IT teams.
D. Software development may require more detailed specifications.

B is the correct answer.

Justification:
A. Higher telecommunications costs are more manageable than privacy laws.
B. Privacy laws prohibiting the cross-border flow of personally identifiable information (PII) may make it impossible to locate a data warehouse containing customer information in another country.
C. Time zone differences are more manageable than privacy laws.
D. Typically, software development requires more detailed specifications when dealing with offshore operations.

R3-31 Which of the following is the **MOST** important factor when designing IS controls in a complex environment?

A. Development methodologies
B. Scalability of the solution
C. Technical platform interfaces
D. Stakeholder requirements

D is the correct answer.

Justification:
A. Development methodologies are taken into consideration when designing IS controls to support stakeholder requirements.
B. Scalability of the solution is taken into consideration when designing IS controls to support stakeholder requirements.
C. Technical platform interfaces are taken into consideration when designing IS controls to support stakeholder requirements.
D. The most important factor when designing IS controls is that they advance the interests of the business by addressing stakeholder requirements.

R3-32 A global enterprise that is subject to regulation by multiple governmental jurisdictions with differing requirements should:

A. bring all locations into conformity with the aggregate requirements of all governmental jurisdictions.
B. bring all locations into conformity with a generally accepted set of industry best practices.
C. establish a baseline standard incorporating those requirements that all jurisdictions have in common.
D. establish baseline standards for all locations and add supplemental standards as required.

D is the correct answer.

Justification:
A. Seeking a lowest common denominator of requirements may cause certain locations to fail regulatory compliance.
B. Just using industry best practices may cause certain locations to fail regulatory compliance.
C. Forcing all locations to be in compliance with the regulations places an undue burden on those locations.
D. It is more efficient to establish a baseline standard and then develop additional standards for locations that must meet specific requirements.

R3-33 The person responsible for ensuring that information is classified is the:

A. security manager.
B. technology group.
C. data owner.
D. senior management.

C is the correct answer.

Justification:
A. The security manager is responsible for applying security protection relative to the level of classification specified by the owner.
B. The technology group is delegated custody of the data by the data owner, but the group does not classify the information.
C. The data owner is responsible for applying the proper classification to the data.
D. Senior management is ultimately responsible for the enterprise.

R3-34 When transmitting personal information across networks, there **MUST** be adequate controls over:

A. encrypting the personal information.
B. obtaining consent to transfer personal information.
C. ensuring the privacy of the personal information.
D. change management.

C is the correct answer.

Justification:
A. Encryption is a method of achieving the actual control, but controls over the devices may not ensure adequate privacy protection. Therefore, encryption is a partial answer.
B. Consent is one of the protections that are frequently, but not always, required.
C. Privacy protection is necessary to ensure that the receiving party has the appropriate level of protection of personal data.
D. Change management is a core control that ensures that the privacy protections, encryption settings and consent processes are implemented as management intended; however, it will not directly address the privacy of the individuals.

R3-35 Which of the following **BEST** addresses the risk of data leakage?

A. Incident response procedures
B. File backup procedures
C. Acceptable use policies (AUPs)
D. Database integrity checks

C is the correct answer.

Justification:
A. Confidentiality of information is not addressed by this choice.
B. Confidentiality of information is not addressed by this choice.
C. Acceptable use policies (AUPs) are the best measure for preventing the unauthorized disclosure of confidential information.
D. Confidentiality of information is not addressed by this choice.

R3-36 Which of the following devices should be placed within a demilitarized zone (DMZ)?

A. An authentication server
B. A mail relay
C. A firewall
D. A router

B is the correct answer.

Justification:

A. An authentication server, due to its sensitivity, should always be placed on the internal network, never on a demilitarized zone (DMZ) that is subject to compromise.
B. A mail relay should normally be placed within a DMZ to shield the internal network.
C. Firewalls may bridge a DMZ to another network, but do not technically reside within the DMZ network segment.
D. Routers may bridge a DMZ to another network, but do not technically reside within the DMZ network segment.

R3-37 Which of the following controls within the user provision process **BEST** enhances the removal of system access for contractors and other temporary users when it is no longer required?

A. Log all account usage and send it to their manager.
B. Establish predetermined, automatic expiration dates.
C. Ensure that each individual has signed a security acknowledgement.
D. Require managers to email security when the user leaves.

B is the correct answer.

Justification:

A. Logging is a detective control and, thus, is not as effective as the protective control of preexpiring user accounts.
B. Predetermined expiration dates are the most effective means of removing systems access for temporary users.
C. Requiring each individual to sign a security acknowledgement has little effect in this case.
A. Managers cannot be relied on to promptly send in termination notices.

R3-38 Which of the following **BEST** provides message integrity, sender identity authentication and nonrepudiation?

A. Symmetric cryptography
B. Message hashing
C. Message authentication code
D. Public key infrastructure (PKI)

D is the correct answer.

Justification:

A. Symmetric cryptography provides confidentiality.
B. Hashing can provide integrity and confidentiality.
C. Message authentication codes provide integrity.
D. Public key infrastructure (PKI) combines public key encryption with a trusted third party to publish and revoke digital certificates that contain the public key of the sender. Senders can digitally sign a message with their private key and attach their digital certificate (provided by the trusted third party). These characteristics allow senders to provide authentication, integrity validation and nonrepudiation.

R3-39 Which of the following will **BEST** prevent external security attacks?

A. Securing and analyzing system access logs
B. Network address translation
C. Background checks for temporary employees
D. Static Internet protocol (IP) addressing

B is the correct answer.

Justification:
A. Securing and analyzing system access logs is a detective control.
B. Network address translation is helpful by having internal addresses that are nonroutable.
C. Background checks of temporary employees are more likely to prevent an attack launched from within the enterprise.
D. Static Internet protocol (IP) addressing does little to prevent an attack.

R3-40 Which of the following is the **BEST** control for securing data on mobile universal serial bus (USB) drives?

A. Requiring authentication when using USB devices
B. Prohibiting employees from copying data to USB devices
C. Encrypting USB devices
D. Limiting the use of USB devices

C is the correct answer..

Justification:
A. Authentication is the act of verifying identity of a user, system, service, etc. Authentication is generally coupled with identification and the delivery of access privileges; by itself, it is not a strong security measure.
B. Prohibiting employees from copying data to universal serial bus (USB) devices is a directive control that is challenging to enforce. It is not a strong security measure.
C. Encryption provides the most effective protection of data on mobile devices.
D. Limiting the use of USB devices is a preventive control that is challenging to enforce. It is not a strong security measure.

R3-41 When configuring a biometric access control system that protects a high-security data center, the system's sensitivity level should be set to:

A. a lower equal error rate (EER).
B. a higher false acceptance rate (FAR).
C. a higher false reject rate (FRR).
D. the crossover error rate exactly.

C is the correct answer.

Justification:

A. When adjusting the sensitivity of a biometric access control system, the values for false acceptance rate (FAR) and false reject rate (FRR) adjust inversely (as indicated in the graph). At one point, the two values intersect and are equal. This condition creates the equal error rate (also called crossover rate), which is a measure of system accuracy.

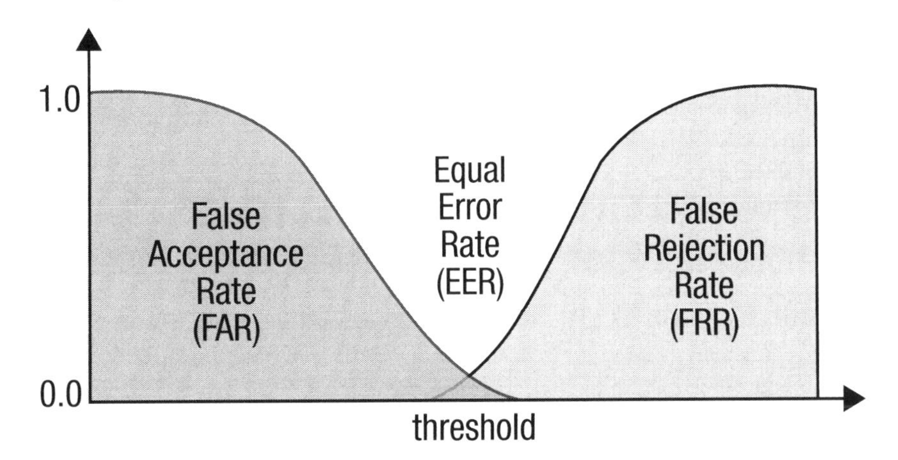

B. A higher FAR is not desirable for a biometric system protecting a high-security data center because it is prone to falsely granting access to unauthorized individuals.

C. Biometric access control systems are not infallible. When tuning the solution for a high-security data center, the sensitivity level should be adjusted to give preference either to an FRR (type I error rate) in which the system will be more prone to falsely reject access to a valid user than falsely granting access to an invalid user. In a very sensitive system, it may be desirable to minimize the number of false accepts—the number of unauthorized people allowed access. To do this, the system is tuned to be more sensitive, which causes the false rejects to increase—the number of authorized people not allowed access.

D. While the crossover error rate is a measure of system accuracy, the best solution for a biometric access control system that protects a high-security data center is to err on the side of false rejects.

R3-42 Which of the following is the **MOST** effective measure to protect data held on mobile computing devices?

A. Protection of data being transmitted
B. Encryption of stored data
C. Power-on passwords
D. Biometric access control

B is the correct answer.

Justification:
A. While protecting data during transmission is important, it does not protect the data stored on the mobile device.
B. Encryption of stored data will help ensure that the actual data cannot be recovered without the encryption key.
C. Power-on passwords do not protect data effectively.
D. Biometric access control does not necessarily protect stored data.

R3-43 Which of the following is **MOST** useful in managing increasingly complex deployments?

A. Policy development
B. A security architecture
C. Senior management support
D. A standards-based approach

B is the correct answer.

Justification:
A. Although policies guide direction, they do not effectively enable complex deployments.
B. Deploying complex security initiatives and integrating a range of diverse projects and activities is more easily managed with the overview and relationships provided by a security architecture.
C. Senior management support is important, yet is insufficient to ensure deployment.
D. Although standards may provide metrics for deployment, they do not effectively enable complex deployments.

R3-44 Business continuity plans (BCPs) should be written and maintained by:

A. the information security and information technology functions.
B. representatives from all functional units.
C. the risk management function.
D. executive management.

B is the correct answer.

Justification:
A. Although the information security and information technology functions areas have primary responsibility for disaster recovery planning, the business continuity plan (BCP) is a primary input representing the business priorities for IT/IS to build, test and maintain disaster recovery plans (DRPs).
B. Business continuity planning is an enterprisewide activity; it is only successful if all business owners collaborate in the development, testing and maintenance of the plan.
C. In many enterprises, risk management may oversee the business continuity program, but they are not in the best position to write or maintain business portions of the plan.
D. Executive management are responsible for assuring that the appropriate planning is performed, the plan is viable and understanding their responsibility should the plan be executed.

R3-45 Which of the following is a control designed to prevent segregation of duties (SoD) violations?

A. Enabling IT audit trails
B. Implementing two-way authentication
C. Reporting access log violations
D. Implementing role-based access

D is the correct answer.

Justification:

A. This choice is not correct because IT audits are detective controls and cannot ensure prevention of segregation of duties (SoD) violations.
B. This choice is not correct since two-way authentication ensures that client and server authenticate each other and does not help prevent SoD violations.
C. Reporting access log violations is a detective control and cannot ensure prevention of SoD violations.
D. Implementing role-based access is a preventive method to mitigate SoD violations. All access levels can be adjusted according to the current role of the user, thus avoiding approvals of self-initiated transactions.

R3-46 System backup and restore procedures can **BEST** be classified as:

A. Technical controls
B. Detective controls
C. Corrective controls
D. Deterrent controls

C is the correct answer.

Justification:

A. Technical controls are safeguards incorporated into computer hardware, software or firmware. Operational procedures are nontechnical controls.
B. Detective controls help identify and escalate violations or attempted violations of security policy; examples include audit trails, intrusion detection tools and checksums.
C. Corrective controls remediate vulnerabilities. If a system suffers harm so extensive that processing cannot continue, backup restore procedures enable that system to be recovered. This is a corrective measure that remediates the vulnerability of that system.
D. Deterrent controls provide warnings that can discourage potential compromise; examples include warning banners or login screens.

R3-47 Which of the following system development life cycle (SDLC) stages is **MOST** suitable for incorporating internal controls?

A. Development
B. Testing
C. Implementation
D. Design

D is the correct answer.

Justification:

A. Internal control requirements should be incorporated during development; however, unless the team already started incorporating internal controls during the preceding design phase, the project may incur a rework cost and is likely to affect project deliverables, project cost and the project time line.
B. Incorporating internal control requirements as late as the testing stage is likely to adversely affect project deliverables, project cost and the project time line.
C. Incorporating internal control requirements as late as the implementation stage is too late and may pose significant risk to the enterprise.
D. Internal controls should be incorporated in the new system development at the earliest stage possible (i.e., at the design stage).

R3-48 An enterprise has outsourced personnel data processing to a supplier, and a regulatory violation occurs during processing. Who will be held legally responsible?

A. The supplier, because it has the operational responsibility
B. The enterprise, because it owns the data
C. The enterprise and the supplier
D. The supplier, because it did not comply with the contract

B is the correct answer.

Justification:

A. The supplier has the operational responsibility pursuant to the contractual terms, but the regulatory authority will hold the customer responsible.
B. The enterprise retains responsibility for the management of, and adherence to, policies, procedures and regulatory requirements. If the supplier fails to provide appropriate controls and/or performance based on the contract terms, the enterprise may have legal recourse. However, the regulatory authority will generally hold the enterprise responsible for failure to comply with regulations, including any penalties that may result.
C. From the point of view of the regulatory authority the enterprise is legally responsible; in other words, the enterprise may be litigated and/or penalized for not fulfilling the contractual terms.
D. The supplier has the operational responsibility pursuant to the contractual terms, but the regulatory authority will hold the enterprise responsible.

R3-49 Which of the following provides the formal authorization on user access?

A. Database administrator
B. Data owner
C. Process owner
D. Data custodian

B is the correct answer.

Justification:

A. The database administrator is responsible for overall database maintenance, support and performance and may grant access to data within the database once the data owner has approved the access request.
B. The data owner provides the formal authorization to provide access to any user request.
C. The process owner is responsible for a specific business process.
D. The data custodian is responsible for the safe custody, transport and storage of data and implementation of business rules, such as granting access to data, once the data owner has approved the access request.

R3-50 To determine the level of protection required for securing personally identifiable information, a risk practitioner should **PRIMARILY** consider the information:

A. source.
B. cost.
C. sensitivity.
D. validity.

C is the correct answer.

Justification:

A. The level of protection required is independent of the source of information and has more to do with the validity and information reliability than with protection.
B. The cost incurred to procure the information can partly determine the level of protection and in some way strengthens the information sensitivity factor, but cannot be the sole factor to determine protection level.
C. Sensitivity of the information is the correct answer because the sensitive nature of the information takes precedence over source, cost or reliability.
D. Validity of information does not dictate level of protection because protection is based on the content categorization.

R3-51 Risk assessments are **MOST** effective in a software development organization when they are performed:

A. before system development begins.
B. during system deployment.
C. during each stage of the system development life cycle (SDLC).
D. before developing a business case.

C is the correct answer.

Justification:

A. Performing a risk assessment before system development does not reveal any of the vulnerabilities introduced during development.
B. Performing a risk assessment at system deployment is not cost effective.
C. Performing risk assessments at each stage of the system development life cycle (SDLC) is the most cost-effective way because it ensures that flaws are caught as soon as they occur.
D. Performing a risk assessment before developing a business case does not reveal any of the vulnerabilities found during the SDLC.

R3-52 Security technologies should be selected **PRIMARILY** on the basis of their:

A. evaluation in security publications.
B. compliance with industry standards.
C. ability to mitigate risk to organizational objectives.
D. cost compared to the enterprise's IT budget.

C is the correct answer.

Justification:

A. Evaluation in security publications is a valuable reference point when selecting a security technology, yet is secondary to the technology's ability to mitigate risk to the enterprise.
B. Compliance with industry standards may be an important aspect of selecting a security technology, but is secondary to the technology's ability to mitigate risk to the enterprise.
C. The most fundamental evaluation criterion for the selection of security technology is its ability to reduce risk.
D. While the cost of the technology in comparison to the budget is an important aspect for the selection of a suitable technology, it is secondary to the technology's ability to mitigate risk to the enterprise.

R3-53 Which of the following groups would be the **MOST** effective in managing and executing an organization's risk program?

A. Midlevel management
B. Senior management
C. Frontline employees
D. The incident response team

A is the correct answer.

Justification:

A. Midlevel management staff are the best to manage and execute an organization's risk management program because they are the most centrally located within the organizational hierarchy and they combine a sufficient breadth of influence with adequate proximity to day-to-day operations.

B. Senior management staff are at too high a level to manage and execute the program, but their support is essential.

C. Frontline employees do not hold enough power and influence to manage and execute the program.

D. The incident response team may manage and execute a small portion of the program (incident response), but not the entire risk management program.

R3-54 Strong authentication is:

A. an authentication technique formally approved by a standardization organization.
B. the simultaneous use of several authentication techniques, e.g., password and badge.
C. an authentication system that makes use of cryptography.
D. an authentication system that uses biometric data to identify a person, e.g., a fingerprint.

B is the correct answer.

Justification:

A. The use of a standardized authentication technique in itself does not imply strong authentication.

B. Authentication is the process of proving to someone that you are who you say you are—a guarantee of the sender's identity or origin. Because a third party vouches for the sender's identity, the recipient can rely on the authenticity of any transaction or message signed by that user. *Strong authentication* requires both something you *know* AND either something you have or *are*.
Three classic methods of authentication are:
• Something you *know*—passwords, the combination to a safe
• Something you *have*—keys, tokens, badges
• Something you *are*—physical traits, such as fingerprints, signature, iris pattern, keystroke patterns

C. Cryptography is the practice and study of hiding information and—in relation to authentication—is mostly used to protect information that may be misused by a third party, such as passwords.

D. Biometrics consists of methods for uniquely recognizing humans based on one or more intrinsic physical or behavioral traits; it can help strengthen the authentication processes, but in itself does not imply strong authentication.

R3-55 The board of directors of a one-year-old start-up company has asked their chief information officer (CIO) to create all of the enterprise's IT policies and procedures, which will be managed and approved by the IT steering committee. The IT steering committee will make all of the IT decisions for the enterprise, including those related to the technology budget. The IT steering committee will be **BEST** represented by:

A. members of the executive board.
B. high-level members of the IT department.
C. IT experts from outside of the enterprise.
D. key members from each department.

D is the correct answer.

Justification:
A. If the steering committee is comprised of only the executive board, then it is likely that all of the goals will be high level and it is impractical for daily IT decisions to be made by the executive board.
B. If the steering committee is comprised of only the IT department, then it is likely that business objectives will be ignored in favor of technical best practices.
C. If the steering committee is comprised of only experts from outside of the enterprise, then business objectives will likely be ignored.
D. The IT steering committee should be comprised of individuals from each department to ensure that the entire enterprise is represented and that all business objectives are more likely to be met.

R3-56 Information security procedures should:

A. be updated frequently as new software is released.
B. underline the importance of security governance.
C. define the allowable limits of behavior.
D. describe security baselines for each platform.

A is the correct answer.

Justification:
A. Often, security procedures have to change frequently to keep up with changes in software. Because a procedure is a how-to document, it must be kept up-to-date with frequent changes in software.
B. High-level objectives of an enterprise, such as security governance, are normally addressed in a security policy.
C. Security policies define behavioral limits and are generally not updated as frequently as procedures.
D. Security standards define platform baselines; however, they do not provide the detail on how to apply the security baseline and are generally not updated as frequently as procedures.

R3-57 Security administration efforts are **BEST** reduced through the deployment of:

A. access control lists (ACLs).
B. discretionary access controls (DACs).
C. mandatory access controls (MACs).
D. role-based access controls (RBACs).

D is the correct answer.

Justification:

A. Access control lists (ACLs) fall in the category of discretionary access controls in that they assign permissions to specific operations with meaning in the enterprise, and not to the low-level data objects. ACLs can never be as fine-tuned as RBACs and need frequent modification.
B. The traditional discretionary access control (DAC) model is based on resource ownership and access to resources is based on user identity. DAC requires changes to a user access profile every time that the identity changes.
C. In a mandatory access control (MAC) model, access to resources is based on an object's security level, while users are granted security clearance. Only administrators can modify an object's security label or a user's security clearance, and this increases the administrative overhead cost.
D. Role-based access controls (RBACs) tie individuals to specific roles. The use of roles, hierarchies and constraints to organize privileges reduces the security administration effort when individuals change positions. RBACs are also known as nondiscretionary access controls.

R3-58 Which of the following is the **BEST** approach when malicious code from a spear phishing attack resides on the network and the finance department is concerned that scanning the network will slow down work and delay quarter-end reporting?

A. Instruct finance to finalize quarter-end reporting, and then perform a scan of the entire network.
B. Block all outgoing traffic to avoid outbound communication to the expecting command host.
C. Scan network devices that are not supporting financial reporting, and then scan the critical finance drives at night.
D. Perform a staff survey and ask staff to report if they are aware of the enterprise being a target of a spear phishing attack.

C is the correct answer.

Justification:

A. Instructing finance to finalize quarter-end reporting and not performing scanning until it is complete is not the most efficient approach to deal with a potential incident.
B. Blocking all outgoing traffic to avoid outbound communication to the expecting command host will also block justified outbound business communication and is not the most effective approach to deal with a potential incident.
C. **Implementing an incremental scanning approach helps confirm the potential risk while allowing the business unit responsible for financial reporting to conduct their operations with minimal interference.**
D. Asking staff if they are aware of being a target of a spear phishing attack is useless because spear phishing attacks are specifically designed to look like authentic email messages and users are not likely to be aware that they have been victimized.

R3-59 Which of the following is the **BEST** option to ensure that corrective actions are taken after a risk assessment is performed?

A. Conduct a follow-up review.
B. Interview staff member(s) responsible for implementing the corrective action.
C. Ensure that an organizational executive documents that the corrective action was taken.
D. Run a monthly report and verify that the corrective action was taken.

A is the correct answer.

Justification:
A. Conducting a follow-up review is correct because it is the only option that ensures that the corrective action was taken.
B. Interviewing the staff member(s) is incorrect because there is no concrete proof that the action was taken.
C. Documenting that a corrective action was taken is incorrect because it does not ensure that the action was taken, but it is a good step to take once a follow-up audit is conducted.
D. A monthly report may not be specific enough to ensure that the corrective action was taken.

R3-60 Which of the following **BEST** ensures that appropriate mitigation occurs on identified information systems vulnerabilities?

A. Presenting root cause analysis to the management of the organization
B. Implementing software to input the action points
C. Incorporating the findings into the annual report to shareholders
D. Assigning action plans with deadlines to responsible personnel

D is the correct answer.

Justification:
A. Presenting findings to management will increase management awareness; however, it does not ensure that action will be taken by the staff.
B. Software can help in monitoring the progress of mitigations, but it will not ensure completeness.
C. Reporting to shareholders does not ensure that the mitigation will be completed.
D. Assigning mitigation to personnel establishes responsibility for its completion within the deadline.

R3-61 Which of the following **BEST** ensures that information systems control deficiencies are appropriately remediated?

A. A risk mitigation plan
B. Risk reassessment
C. Control risk reevaluation
D. Countermeasure analysis

A is the correct answer.

Justification:
A. Once risk is identified due to current IS control deficiencies, a risk mitigation plan will have the set of controls with a detailed plan, including countermeasures that can best help in risk remediation to an appropriate level.
B. Risk reassessment is required when there are major changes to the risk environment; it is usually performed after a period of time as defined by management.
C. Control risk reevaluation helps in further validation of IS control deficiencies, but does not ensure that these deficiencies are actually remediated.
D. Countermeasure analysis is targeted toward countermeasures and may provide some information on deficiencies, but does not ensure that IS control deficiencies are remediated.

R3-62 Which organizational function is accountable for risk policies, guidelines and standards?

A. Operations
B. IT
C. Management
D. Legal

C is the correct answer.

Justification:

A. Operations controls, implemented in accordance with a base set of requirements (e.g., technical controls) and good industry practices, are used to correct operational deficiencies that could be exercised by potential threat sources. To ensure consistency and uniformity in security operations, step-by-step procedures and methods for implementing operational controls must be clearly defined, documented and maintained.
B. IT security controls for risk mitigation can be configured to protect against given types of threats. These controls may range from simple to complex measures and usually involve system architectures; engineering disciplines; and security packages with a mix of hardware, software and firmware. All of these measures should work together to secure critical and sensitive data, information and IT system functions.
C. Management security controls, in conjunction with technical and operational controls, are implemented to manage and reduce the risk of loss and to protect an enterprise's mission. Management controls focus on the stipulation of information protection policy, guidelines and standards, which are carried out through operational procedures to fulfill the enterprise's goals and missions.
D. Legal does not own risk policies, guidelines and standards.

R3-63 The risk action plan **MUST** include an appropriate resolution, a date for completion and:

A. responsible personnel.
B. mitigating factors.
C. likelihood of occurrence.
D. cost of completion.

A is the correct answer.

Justification:

A. Risk response activities must be assigned to the responsible person or group; if this is not included, it will be unclear who will implement the countermeasure.
B. Mitigating factors can be included, but is not as important as responsible personnel.
C. Compensating controls can be included, but are not as important as responsible personnel.
D. Cost for completion is an optional field and is not necessary.

R3-64 Risk response should focus on which of the following?

A. Destruction of obsolete computer equipment
B. Theft of a smart phone from an office
C. Sanitization and reuse of a flash drive
D. Employee deletion of a file

B is the correct answer.

Justification:
A. Destruction of obsolete computer equipment is an example of an operational activity.
B. Disposal of data should be addressed in the IT operations function. Risk response should focus on nonoperational data disposal (loss or theft). Theft of a smart phone is an example of a risk that should be addressed by an appropriate response such as a remote wipe.
C. Sanitization of a flash drive is an example of an operational activity.
D. Deletion of a file is an example of an operational activity.

R3-65 Which of the following risk response options is **MOST** likely to increase the liability of the enterprise?

A. Risk acceptance
B. Risk reduction
C. Risk transfer
D. Risk avoidance

A is the correct answer.

Justification:
A. An enterprise may choose to accept risk without knowing the correct level of risk that is being accepted; this may result in accusations of negligence.
B. Risk reduction indicates an attempt to reduce the risk level. It may not be as effective as intended, but is not likely to increase the level of risk.
C. Risk transfer allocates a portion of risk to another party (e.g., insurance).
D. Risk avoidance will terminate a process that is considered to have an unacceptable level of risk that cannot be mitigated economically.

R3-66 Which of the following is the **BEST** reason an enterprise would decide not to reduce an identified risk?

A. There is no regulatory requirement to reduce the risk.
B. The inherent risk of the related business process is low.
C. The potential gain outweighs the risk.
D. The cost of reducing the risk exceeds the budget.

C is the correct answer.

Justification:
A. Regulatory requirements are not the only risk factors affecting an enterprise's decision to reduce risk; other factors may include reputational damage, financial repercussions and others.
B. The low inherent risk of one business process may affect other more high-risk business processes.
C. Risk is not the main driver for the business/enterprise decision process. The business will accept the risk when it is determined that the potential opportunities may yield a higher return in revenue and/or gain in market share compared to risk.
D. Enterprises may choose to reduce a risk even when the cost exceeds the budget, such as when the cost of reducing the risk is lower than the projected impact of the risk.

R3-67 An enterprise decides to address risk associated with an IT project by outsourcing part of the IT activities to a third party with a specialized skill set. In relation to the project itself, this is an example of:

A. risk transfer.
B. risk avoidance.
C. risk acceptance.
D. risk mitigation.

D is the correct answer.

Justification:

A. Outsourcing part of an activity is not risk transfer. Risk transfer assigns risk to another enterprise, usually through the purchase of an insurance policy or by outsourcing the service.
B. Outsourcing part of an activity is not risk avoidance. Risk avoidance is the process for systematically steering away from a specific risk, generally by not engaging in a specific activity, such as e-commerce or cloud computing. Risk avoidance generally also affects the potential opportunity offered by engaging in the activity.
C. Outsourcing part of an activity is not risk acceptance. Risk acceptance means that the enterprise makes an educated decision not to take action relative to a particular risk and accepts loss when/if it occurs.
D. Outsourcing part of an activity in itself does not transfer risk; the risk remains with the enterprise. However, when specific activities are outsourced to an entity with a specialized skill set, the inherent risk of the activity is reduced.

R3-68 Which of the following **BEST** helps to respond to risk in a cost-effective manner?

A. Prioritizing and addressing risk according to the risk management strategy
B. Mitigating risk on the basis of risk likelihood and magnitude of impact
C. Performing countermeasure analysis for each of the controls deployed
D. Selecting controls that are at zero or near-zero costs

A is the correct answer.

Justification:

A. If risk is prioritized and addressed in line with the risk treatment strategy, it balances the costs and benefits of managing the IT risk.
B. Mitigating risk on the basis of risk likelihood and magnitude of impact is difficult because there can be multiple occurrences of risk where the product of likelihood multiplied by impact is very similar. Without prioritization; risk treatment will not be optimized.
C. Performing countermeasure analysis for each control that is deployed does not help because there may not be a countermeasure for every control. Even if there were, it would not help in the risk response.
D. Selecting controls that are at zero or near-zero costs may reduce the costs in general, but the controls themselves may not be effective.

R3-69 Which of the following **BEST** assists in the development of the risk profile?

A. The presence of preventive and detective controls
B. Inherent risk and detection risk
C. Cost-benefit analysis of controls
D. Likelihood and impact of risk

D is the correct answer.

Justification:

A. Preventive and detective controls by themselves do not help in the development of the risk profile.
B. Inherent risk and detection risk are components of the total risk and by themselves do not help in developing the risk profile. Inherent risk is the risk level or exposure without taking into account the actions that management has taken or might take (e.g., implementing controls). Detection risk is the risk that material errors or misstatements have occurred and will not be detected by the IS auditor.
C. Cost-benefit analysis of controls helps in selecting controls for a given risk, but not in development of the risk profile itself.
D. Likelihood and impact of risk in itself helps in the development of the risk profile.

R3-70 Which of the following can **BEST** be used as a basis for recommending a data leak prevention (DLP) device as a security control?

A. Benchmarking with peers on DLP deployment
B. A business case for DLP to protect data
C. Evaluation report of popular DLP solutions
D. DLP scenario in risk register

B is the correct answer.

Justification:

A. Benchmarking with peers does not help because peers will have a different risk environment and culture that do not directly apply to one's own enterprise.
B. A business case with costs vs. benefits provides the business reasoning why the data leak prevention (DLP) solution addresses the risk and explains how the risk losses could be reduced if the data were leaked.
C. While this is important information, the risk practitioner still needs to analyze the report in the context of the enterprise and recommend the appropriate solution.
D. Merely identifying data leakage in the risk register does not help in justifying the solution.

R3-71 Which of the following is **BEST** performed for business continuity management to meet external stakeholder expectations?

A. Prioritize applications based on business criticality.
B. Ensure that backup data are available to be restored.
C. Disclose the crisis management strategy statement.
D. Obtain risk assessment by an independent party.

A is the correct answer.

Justification:

A. External parties (such as customers) expect that their information assets are secured. To meet this goal, it is strategically important to prioritize applications based on business criticality. With this approach, their expectations can be maximized with the use of limited resources.

B. Ensuring that backup data can readily be restored is a fundamental requirement for business continuity. However, it is less likely that external stakeholders raise this point as a top agenda item.

C. External customers look for various means to determine that the companies in which they invest/trust are capable of recovery from an unanticipated event. Based on industry/country/regulatory requirements, some enterprises do disclose a "public" crisis management strategy statement, but this is not the best answer.

D. Obtaining third-party assessment is quite important. However, this type of assessment does not necessarily guarantee that it will meet external stakeholder expectations.

R3-72 When the risk related to a specific business process is greater than the potential opportunity, the **BEST** risk response is:

A. transfer.
B. acceptance.
C. mitigation.
D. avoidance.

D is the correct answer.

Justification:

A. Risk transfer is the process of assigning risk to another enterprise, usually through the purchase of an insurance policy or by outsourcing the service.

B. Risk acceptance means that no action is taken relative to a particular risk, and loss is accepted when/if it occurs.

C. Risk mitigation is the management of risk through the use of countermeasures and controls.

D. Risk avoidance is the process for systematically avoiding risk, constituting one approach to managing risk.

R3-73 Which of the following should management use to allocate resources for risk response?

A. Audit report findings
B. Penetration test results
C. Risk analysis results
D. Vulnerability test results

C is the correct answer.

Justification:
A. An audit report provides recommendation and remediation areas.
B. Penetration test results help identify vulnerabilities.
C. Risk analysis results provide a basis for prioritization of risk responses and the allocation of resources.
D. Vulnerability test results provide an enterprise with a list of known vulnerabilities for the systems that have been assessed. They do not take "control in depth" considerations into account and are not a meaningful tool for determining the allocation of risk response resources.

R3-74 Which of the following **BEST** identifies changes in an enterprise's risk profile?

A. The risk register
B. Risk classification
C. Changes in risk indicator thresholds
D. Updates to the control inventory

A is the correct answer.

Justification:
A. The risk register is the central document to identify changes in an enterprise's risk profile.
B. Risk classification helps prioritize risk for an effective risk response, but does not capture changes in the business environment.
C. Changes in risk indicator thresholds affect the times at which a risk event is realized. However, such a change does not capture changes in the business environment.
D. Updates to the control inventory are an important input into the risk management process because they are an important internal environmental risk factor. Other risk factors include external environment, internal capacity, IT capacity and others.

R3-75 Which of the following **BEST** identifies controls addressing risk related to cloud computing?

A. Data encryption, tenant isolation, controlled change management
B. Data encryption, customizing the application template, creating and importing custom widgets
C. Selecting an open standards-based technology, data encryption, tenant isolation
D. Tenant isolation, controlled change management, creating and importing custom widgets

A is the correct answer.

Justification:
A. One of the baseline controls—encryption—enables keeping data separate from other tenants. Tenant isolation, as opposed to comingling tenants, is a basic way of keeping data separate from multiple tenants. Having a controlled change management process ensures no surprises from either the tenant or the vendor, and that all changes are well planned and tenant dependencies are mapped to underlying resources and services.
B. Customizing the application template and importing custom widgets are application or software activities that do not specifically relate to data.
C. Selecting an open standards-based technology is not specifically related to controls addressing data risk.
D. As in choice B, importing custom widgets is an application or software activity that does not specifically relate to data.

R3-76 Prior to releasing an operating system security patch into production, a leading practice is to have the patch:

A. applied simultaneously to all systems.
B. procured from an approved vendor.
C. tested in a preproduction test environment.
D. approved by business stakeholders.

C is the correct answer.

Justification:

A. Although it is important to ensure that all devices are adequately patched in a timely fashion, patches should be released in phases to the related systems, starting with critical systems or the most vulnerable systems and then working down the priority chain.
B. Procuring patches from an approved vendor runs the risk that the patches might not work in the environment.
C. When a change goes into production, the most important practice is to ensure that testing has been completed. In the case of a security patch, testing is essential because an untested security patch may cause serious business disruptions.
D. Having the business stakeholders sign off on the patch release to production is not required because most of the operating system (OS) patches are technical in nature and are released after due process by the operations team.

R3-77 Which of the following helps ensure that the cost is justifiable when selecting an IT control?

A. The investment is within budget.
B. The risk likelihood and its impact are reflected.
C. The net present value (NPV) is high.
D. Open source technology is used.

B is the correct answer.

Justification:

A. The fact that the cost of a control is within budget does not necessarily justify the cost of a control. The cost of a control should be less than the projected benefit of the control.
B. While other factors may be relevant, the total cost of ownership of a control should not exceed the projected likelihood times the impact of the risk it is intended to mitigate.
C. The net present value (NPV) is calculated by using an after-tax discount rate of an investment and a series of expected incremental cash outflows (the initial investment and operational costs) and cash inflows (cost savings or revenues) that occur at regular periods during the life cycle of the investment. It does not justify the cost of the control because it does not relate the cost to the expected benefit.
D. While open source technology is generally a low-cost option, the low cost itself does not justify the cost of the control.

R3-78 The **PRIMARY** purpose of providing built-in audit trails in applications is to:

A. support e-discovery.
B. collect information for auditors.
C. enable troubleshooting.
D. establish accountability.

D is the correct answer.

Justification:

A. Although audit trails are a part of e-discovery, the primary purpose of an audit trail is still to establish accountability of who did what and when.
B. Even without audit trails, auditors can always use others tools and methods to verify controls.
C. Audit trails are used during troubleshooting to determine what happened before a given event; however, the primary day-to-day function is establishing accountabilities for actions taken.
D. Audit trails will record the various events that are processed as part of the complete transaction and therefore establish accountability for processed transaction because no one can deny the facts regarding the processed transactions.

R3-79 Which of the following **BEST** protects the confidentiality of data being transmitted over a network?

A. Data are encapsulated in data packets with authentication headers.
B. A digital hash is appended to all messages sent over the network.
C. Network devices are hardened in compliance with corporate standards.
D. Fiber-optic cables are used instead of copper cables.

A is the correct answer.

Justification:

A. Data are encapsulated in data packets with authentication headers that will protect confidentiality even if there is man-in-the-middle attack or interception of the data by other means.
B. A digital hash appended to the messages will only ensure data integrity.
C. Network device hardening will only protect data at the end points and not during transmission.
D. While fiber-optic cables have several advantages over copper cables, the choice of cabling itself does not sufficiently address the confidentiality of data being transmitted over a network.

R3-80 Which of the following is the **MOST** significant risk associated with handling credit card data through a web application?

A. Displaying both the first six and last four digits of the credit card, thus exposing sensitive information
B. Allowing the transmission of credit card data over the Internet using an insecure channel such as Secure Sockets Layer (SSL) protocol or Transport Layer Security (TLS) protocol
C. Failure to store credit card data in a secure area segregated from the demilitarized zone (DMZ)
D. Installation of network devices with default access settings disabled or inoperable

C is the correct answer.

Justification:

A. The Payment Card Industry Data Security Standard (PCI DSS) allows the display of both the first six and last four digits; only the six digits in the middle must be protected.
B. Secure Sockets Layer (SSL) and Transport Layer Security (TLS) are secure protocols commonly used to transmit sensitive data over the Internet.
C. Failure to store credit card data in a secure area segregated from the demilitarized zone (DMZ) is one of the most common and serious flaws in a secure architecture.
D. The default setting must be changed on all network devices that will process credit card data.

R3-81 Which of the following is the **MOST** important consideration when developing a record retention policy?

A. Delete, as quickly as practical, all data that are not required.
B. Retain data only as long as necessary for business or regulatory requirements.
C. Keep data to ensure future availability.
D. Archive old data without encryption as quickly as practical.

B is the correct answer.

Justification:
A. Deleting superfluous data is a process that may be called for by the policy. It is not a consideration when developing a records retention policy.
B. Good practice states that data should be kept only as long as required by business or regulation requirements.
C. It is better not to keep any data longer than necessary because the loss of old data may still pose a serious risk to the enterprise.
D. Old data should still be encrypted (and perhaps reencrypted with a stronger algorithm) if they are being retained by the enterprise.

R3-82 An enterprise is implementing controls to protect its product price list from being exposed to unauthorized individuals. The internal control requirements will come from:

A. the risk management team.
B. internal audit.
C. IT management.
D. process owners.

D is the correct answer.

Justification:
A. The risk management team will be involved in the control risk review process.
B. Internal audit will measure the effectiveness of the implemented internal control.
C. IT management will be involved in the IT governance and policies and procedures, but not with business-specific internal controls.
D. Process owners will provide the internal control requirements based on the business needs and objectives.

R3-83 Which of the following is **MOST** important when mitigating or managing risk?

A. Vulnerability assessment results
B. A business impact analysis (BIA)
C. The risk tolerance level
D. A security controls framework

C is the correct answer.

Justification:

A. Vulnerability assessments provide a view of an enterprise's current control environment, specifically where controls are weak or lacking. Unlike a business impact analysis (BIA), they do not tie vulnerabilities to the enterprise's threat landscape or help determine the impact of adverse events affecting specific business processes.

B. A BIA is the process to determine the impact of losing the support of any resource. The BIA assessment study will establish the escalation of that loss over time. It is predicated on the fact that senior management, when provided reliable data to document the potential impact of a lost resource, can make the appropriate decision.

C. The risk tolerance level (along with risk appetite) determines what kind of risk response an enterprise selects and it needs to be defined in order for an enterprise to appropriately address risk.

D. A framework is a generally accepted, business-process-oriented structure that establishes a common language and enables repeatable business processes. Frameworks generally only describe what needs to be done. They do not provide actionable activities and therefore are not the best solution for determining what risk mitigation activities to pursue.

R3-84 The **MAIN** benefit of information classification is that it helps:

A. determine how information can be further labeled.
B. establish the access control matrices.
C. determine the risk tolerance level.
D. select security measures that are proportional to risk.

D is the correct answer.

Justification:

A. Information labeling consciously identifies the classification levels as determined by the process owners, but this is not the main benefit of classification.

B. An access control matrix is determined based on who should have access to information based on role, and this is not the benefit of classification.

C. The risk tolerance level is determined by current risk level and balances risk that needs to be addressed as part of risk response. Information classification does not help in the process.

D. Based on information classification, information is divided into various buckets of sensitivity, importance and risk so proportional security measures can be designed.

R3-85 Which type of cost incurred is used when leveraging existing network cabling for an IT project?

A. Indirect cost
B. Infrastructure cost
C. Project cost
D. Maintenance cost

A is the correct answer.

Justification:

A. **Indirect costs are often overlooked when calculating the total cost for projects. Full consideration of costs requires attention to both opportunities and indirect costs. For example, the cost of utilizing existing network cabling for a project can be calculated from the amount of new traffic generated or from some other prorating factor.**
B. Infrastructure costs are costs that arise from having property, plant, equipment and a functioning enterprise. IT organizations have an annual budget allocated for infrastructure. The costs can contain network cables, physical servers and data center rentals.
C. Project cost is the total cost of a project, including professional compensation, land costs, furnishings and equipment, and financing.
D. Maintenance costs are periodic costs incurred in activities that preserve an asset's life span, e.g., the costs consist of licensing renewal and hardware parts replacement costs.

R3-86 An enterprise is applying controls to protect its product price list from being exposed to unauthorized staff. These internal controls will include:

A. identification and authentication.
B. authentication and authorization.
C. segregation of duties (SoD) and authorization.
D. availability and confidentiality.

B is the correct answer.

Justification:

A. Identification and authentication are important for confirming the identity of a user; however, both need to be complemented with proper authorization controls to ensure the confidentiality of the price list.
B. **Authentication and authorization are two complementary control objectives that will ensure confidentiality of the price list.**
C. Segregation of duties (SoD) is a control objective that ensures that a single individual is not authorized to execute incompatible activities, such as submitting and approving a change to the price list. SoD and authorization are not the best solution to ensure confidentiality.
D. Availability and confidentiality are business requirements, not controls.

R3-87 During a risk assessment of a start-up company with a bring your own device (BYOD) practice, a risk practitioner notes that the database administrator (DBA) minimizes a social media web site on his/her personal device before running a query of credit card account numbers on a third-party cloud application. The risk practitioner should recommend that the enterprise:

A. develop and deploy an acceptable use policy for BYOD.
B. place a virtualized desktop on each mobile device.
C. blacklist social media web sites for devices inside the demilitarized zone (DMZ).
D. provide the DBA with user awareness training.

B is the correct answer.

Justification:

A. Although it is necessary to have a bring your own device (BYOD) policy before allowing personal devices to attach to a company network, it is a not a preventive control but rather a managerial control.
B. By allowing the BYOD to access the network only via a virtualized desktop client, no data are stored on the device and all the commands entered through the device are actually executed and stored within the enterprise's demilitarized zone (DMZ), network or servers. By using this type of mobile/enterprise architecture, users can be allowed to access the corporate network/data while on a personal device and still be compliant with the enterprise's acceptable use policy.
C. Blacklisting social media web sites or any other application within the DMZ generally does not extend to a personal device attached to the network. It should be noted, however, that there are emerging technologies that can specifically blacklist or whitelist applications on mobile devices.
D. Although every security policy should be accompanied by some level of user awareness training, simply making the database administrator (DBA) aware of potential dangers of using a social media web site and corporate applications simultaneously is not the most effective control.

R3-88 Controls are most effective when they are designed to reduce:

A. threats.
B. likelihood.
C. uncertainty.
D. vulnerabilities.

D is the correct answer.

Justification:

A. A threat is a potential cause of an unwanted incident. Often, threats cannot be reduced, such as in the case of natural disasters.
B. Designing controls to reduce likelihood (of a threat) addresses only one aspect of an event and does not consider a comprehensive approach, including detection and reduction of impact should an adverse event occur.
C. Controls generally do not reduce uncertainty.
D. Controls are most effective when they are designed to reduce vulnerabilities affecting the enterprise. Vulnerabilities can result from external relationships, such as sole-source suppliers.

R3-89 In a large enterprise, system administrators may release critical patches into production without testing. Which of the following would **BEST** mitigate the risk of interoperability issues?

A. Ensure that a reliable system rollback plan is in place.
B. Test the patch on the least critical systems first.
C. Only allow updates to occur after hours.
D. Ensure that patches are approved by the chief information security officer (CISO).

A is the correct answer.

Justification:

A. A reliable system rollback plan will allow the administrators to roll back the patches from the system in case the patches affect the system negatively.
B. Testing the patches on the least critical systems first will not give insight into how the patch will affect all of the applications.
C. Updating the patches after hours will not give insight into how the patch will affect all of the applications, but should be done to decrease the possibility of user downtime.
D. Ensuring that the patches are approved by the chief information security officer (CISO) will not give insight into how the patch will affect all of the applications.

R3-90 Which of the following **BEST** mitigates control risk?

A. Continuous monitoring
B. An effective security awareness program
C. Effective change management procedures
D. Senior management support for control implementation

A is the correct answer.

Justification:

A. Continuous monitoring tests controls that mitigate the risk of the control being less effective over time. A risk assessment will identify when the control is no longer effective and the control will be replaced.
B. An effective security awareness program does not mitigate control risk.
C. Effective change management procedures alone do not mitigate control risk.
D. Senior management support will only assist in implementing a new control, but does not directly mitigate against control risk.

R3-91 During a root cause analysis review of a recent incident it is discovered that the IT department is not tracking any metrics. A risk practitioner should recommend to management that they implement which of the following to reduce the risk?

A. A new help desk system
B. Change management
C. Problem management
D. New reports to track issues

C is the correct answer.

Justification:

A. A new help desk system might help, but an overall problem management process is needed first.
B. Change management is specific to changes and not problems. While it is true that a change can cause a problem, change management is not used to fix a problem.
C. Problem management is part of the Information Technology Information Library (ITIL) and is a process that is used to minimize the impact of problems in an enterprise. Metrics, known errors and incidents are all tracked to minimize problems.
D. New reports might help, but an overall process is needed first.

R3-92 Purchasing insurance is a form of:

A. risk avoidance.
B. risk mitigation.
C. risk acceptance.
D. risk transfer.

D is the correct answer.

Justification:
A. Risk avoidance means that activities of conditions that give rise to risk are discontinued.
B. Risk mitigation is the management of risk through the use of countermeasures and controls. Risk transfer is one form of risk mitigation.
C. Risk acceptance means that no action is taken relative to a particular risk; loss is accepted when/if it occurs.
D. Transferring risk typically involves insurance policies to share the financial risk.

R3-93 Which of the following would **PRIMARILY** help an enterprise select and prioritize risk responses?

A. A cost-benefit analysis of available risk mitigation options
B. The level of acceptable risk per risk appetite
C. The potential to transfer or eliminate the risk
D. The number of controls necessary to reduce the risk

A is the correct answer.

Justification:
A. The selection and priorities of a risk response will consider the cost-benefit of the various risk mitigation options in order to get the highest return on investment (ROI) and reduce the risk to an acceptable level.
B. The level of acceptable risk will not prioritize the risk response, but will indicate whether the risk response is adequate.
C. Risk can be reduced, accepted or transferred. Risk can be transferred (insurance policy) and is an acceptable risk response, but this factor in itself would not help in prioritizing risk responses.
D. The priority for risk mitigation will not be based on the number of controls necessary to reduce the risk, but rather on the implementation of the controls with the greatest cost benefit.

R3-94 Business stakeholders and decision makers reviewing the effectiveness of IT risk responses would **PRIMARILY** validate whether:

A. IT controls eliminate the risk in question.
B. IT controls are continuously monitored.
C. IT controls achieve the desired objectives.
D. IT risk indicators are formally documented.

C is the correct answer.

Justification:
A. Risk cannot be eliminated; the objective is to manage risk to an acceptable level.
B. The continuous monitoring of controls does not necessarily indicate the effectiveness of the control itself.
C. The stakeholders are most interested in whether the control meets the stated objectives.
D. The documentation of IT risk indicators does not demonstrate the effectiveness of the risk response.

R3-95 A risk assessment indicates a risk to the enterprise that exceeds the risk acceptance level set by senior management. What is the **BEST** way to address this risk?

A. Ensure that the risk is quickly brought within acceptable limits, regardless of cost.
B. Recommend mitigating controls if the cost and/or benefit would justify the controls.
C. Recommend that senior management revise the risk acceptance level.
D. Ensure that risk calculations are performed to revalidate the controls.

B is the correct answer.

Justification:

A. Cost needs to be a factor in determining the best way to mitigate a risk. If the cost of the control is higher than the anticipated loss, the risk response is not cost effective.
B. Risk mitigating controls should be implemented based on cost and benefit. Controls are not justified if the cost of the control exceeds the benefit obtained.
C. Revising the risk acceptance level may be desirable in some cases, but this is not the best option.
D. Although it is important that the value of the controls is correct, this will not mitigate the outstanding risk.

R3-96 Which of the following actions is the **BEST** when a critical risk has been identified and the resources to mitigate are not immediately available?

A. Log the risk in the risk register and review it with senior management on a regular basis.
B. Capture the risk in the risk register once resources are available to address the risk.
C. Escalate the risk report to senior management to obtain the resources to mitigate the risk.
D. Review the risk level with senior management and determine whether the risk calculations are correct.

C is the correct answer.

Justification:

A. Because this is a critical risk, logging and reviewing risk on a regular basis would not be a suitable option. It should be escalated to senior management.
B. Because this is a critical risk, capturing the risk once resources are available would not be a suitable option. It should be escalated to senior management.
C. If resources are not available or priorities need to be adjusted, it is important to engage senior management to assist in escalating the remediation.
D. A review of the risk level should already have been performed. This will not resolve the problem with a risk that exceeds the risk acceptance level.

R3-97 Which of the following combinations of factors is the **MOST** important consideration when prioritizing the development of controls and countermeasures?

A. Likelihood and impact
B. Impact and exposure
C. Criticality and sensitivity
D. Value and classification

A is the correct answer.

Justification:

A. The likelihood that a compromise will occur and the impact of that compromise are the two most important factors in determining risk, which in turn drives the development of controls and countermeasures.
B. Impact and exposure combined do not address the likelihood that an incident will occur, so these are not sufficient by themselves.
C. Criticality and sensitivity are the basis for asset classification, but only deal with possible impact, not with likelihood.
D. Valuation is a component of classification and addresses only possible impact.

R3-98 During what stage of the overall risk management process is the cost-benefit analysis **PRIMARILY** performed?

A. During the initial risk assessment
B. During the information asset classification
C. During the definition of the risk profile
D. During the risk response selection

D is the correct answer.

Justification:
A. The cost-benefit analysis is not performed only once, but every time controls need to be decided; this can happen many times over a risk management life cycle.
B. In information asset classification, levels are assigned based on the importance of the asset that needs protection and not based on cost-benefit analysis.
C. The risk profile is defined based on threats and vulnerabilities and what risk needs to be addressed. During this stage no cost-benefit analysis is required.
D. When selecting a risk response, one will identify a range of controls that can mitigate the risk; however, the cost-benefit analysis in this process will help identify the right controls that will address the risk at acceptable levels within the budget.

R3-99 Which of the following approaches **BEST** helps address significant system vulnerabilities that were discovered during a network scan?

A. All significant vulnerabilities must be mitigated in a timely fashion.
B. Treatment should be based on threat, impact and cost considerations.
C. Compensating controls must be implemented for major vulnerabilities.
D. Mitigation options should be proposed for management approval.

B is the correct answer.

Justification:
A. Some vulnerabilities may not have significant impact and may not require mitigation.
B. The treatment should consider the degree of exposure and potential impact and the costs of various treatment options.
C. Compensating controls are considered only when there is a viable threat and impact, and only if the primary control is inadequate.
D. Management approval will depend on a mitigation plan based on threat, impact and cost considerations.

R3-100 What is the **BEST** risk response for risk scenarios where the likelihood is low and financial impact is high?

A. Transfer the risk to a third party.
B. Accept the high cost of protection.
C. Implement detective controls.
D. Implement compensating controls.

A is the correct answer.

Justification:
A. High-impact, low-likelihood situations are typically most cost effectively covered by transferring the risk to a third party, e.g., insurance.
B. Even though financial impact is high, the cost of protection is not necessarily high.
C. A detective control alone does not limit the impact.
D. Implementing compensating controls may be cost prohibitive and is not feasible when the likelihood is already low.

R3-101 Faced with numerous risk, the prioritization of treatment options will be **MOST** effective when based on:

A. the existence of identified threats and vulnerabilities.
B. the likelihood of compromise and subsequent impact.
C. the results of vulnerability scans and exposure.
D. the exposure of corporate assets and operational risk.

B is the correct answer.

Justification:

A. The existence of threats and vulnerabilities affects risk likelihood, but without considering the potential impact, the prioritization of risk treatment will not be effective.
B. The probability of compromise coupled with the likely impact will be the most important considerations for selecting treatment options.
C. Vulnerability scan results provide prioritized input into the decision-making process to remediate technical risk. Vulnerability scans only provide a subset of overall enterprise risk and do not consider the cost to remediate.
D. Exposure of assets and operational risk are factors in determination of prioritization of risk treatment options, but only when used in combination with the cost-benefit of the risk treatment options. Operational risk is a subset of overall risk.

R3-102 Which of the following activities is an example of risk sharing?

A. Moving a function to another department
B. Selling a product or service to another company
C. Deploying redundant firewalls
D. Contracting with a third party

D is the correct answer.

Justification:

A. Moving a function to another department would not share the risk outside of the enterprise's direct responsibility.
B. Selling a product or service to another company would be an example of risk avoidance, not of risk sharing, because the original enterprise no longer has any involvement.
C. Deploying redundant firewalls is risk mitigation, but not risk share/transfer.
D. Contracting with a third party to share the responsibility for supporting activities can provide a form of risk transference as long as it is documented within the outsourcing contract.

R3-103 Which of the following risk response selection parameters results in a decrease in magnitude of an event?

A. Efficiency of response
B. Cost of response
C. Effectiveness of response
D. Capability to implement response

C is the correct answer.

Justification:

A. Efficiency of response is the relative benefit promised by the response in comparison to:
   • Other investments (Investing in risk response measures always competes with other investments.)
   • Other responses (One response may address several risk scenarios while another may not.)
B. Cost of response is the cost of the response to reduce risk to within tolerance levels.
C. Effectiveness of response is the extent to which the response reduces the likelihood and impact.
D. The capability to implement the response will not affect the magnitude of an event.

R3-104 Who grants formal authorization for user access to a protected file?

A. The process owner
B. The system administrator
C. The data owner
D. The security manager

C is the correct answer.

Justification:
A. The process owner is responsible for granting access to a specific business process.
B. The system administrator is the data custodian and is responsible for safe custody, transport and storage of data and grants access to data on approval from the data owner.
C. The data owner grants formal authorization for users to access protected files.
D. The security manager is responsible for enforcing security protection according to the data owner requirements.

R3-105 A business case developed to support risk mitigation efforts for a complex application development project should be retained until:

A. the project is approved.
B. user acceptance of the application.
C. the application is deployed.
D. the application's end of life.

D is the correct answer.

Justification:
A. The business case should be retained even after project approval to justify audit, project review or project scope change.
B. The business case should be retained even after user acceptance to validate the return on investment (ROI).
C. The application may be updated and modified; therefore, the business case should be retained.
D. All documentation related to the system should be updated and retained until the system is no longer in service. The retention may exceed this decommission date if the record retention period declares a longer period.

R3-106 Which of the following tools aids management in determining whether a project should continue based on scope, schedule and cost? Analysis of:

A. earned value management.
B. the function point.
C. the Gantt chart.
D. the program evaluation and review technique (PERT).

A is the correct answer.

Justification:
A. Earned value management takes into consideration scope, schedule and cost in a single integrated system and helps provide accurate forecasts of project performance problems.
B. A function point is a unit of measurement to express the amount of business functionality that an information system provides to a user. It is not suitable for identifying possible overruns on time, budget and deliverables. Function points are factors such as inputs, outputs, inquiries and logical internal sites.
C. Gantt charts focus primarily on tasks and schedule management, but do not consider costs.
D. The program (or project) evaluation and review technique (PERT) is a statistical tool used in project management that is designed to analyze and represent the tasks involved in completing a given project as well as lead and lag times. Similar to the Gantt chart, it does not take cost into consideration.

R3-107 An enterprise has recently implemented a corporate bring your own device (BYOD) policy to reduce the risk of data leakage. Which of the following approaches **MOST** enables the policy to be effective?

A. Obtaining signed acceptance from users on the BYOD policy
B. Educating users on acceptable and unacceptable practices
C. Requiring users to read the BYOD policy and any future updates
D. Clearly stating disciplinary action for noncompliance

B is the correct answer.

Justification:

A. Obtaining sign-off from users does not imply they have read, understand or will follow the details of the policy.
B. While the bring your own device (BYOD) policy is approved and communicated, it would not be effective without proper training of all users.
C. Reading the policy as well as policy updates is important, but it will not be effective unless people are trained on policy requirements.
D. Clearly stating disciplinary action for noncompliance is important, but this is not the best option to ensure that the new policy is effective.

R3-108 In which phase of the system development life cycle (SDLC) should the process to amend the deliverables be defined to prevent the risk of scope creep?

A. Feasibility
B. Development
C. User acceptance
D. Design

A is the correct answer.

Justification:

A. During the feasibility phase (planning or initiation), the process for amending the deliverables is defined, including the authority to approve a change and the process for a change to be submitted.
B. During the development phase, any changes must follow the process defined in the feasibility phase. Uncontrolled changes during this phase would lead to scope creep.
C. During the user acceptance phase, any changes must follow the process defined in the feasibility phase. Uncontrolled changes during this phase would lead to scope creep.
D. During the design phase, any changes must follow the process defined in the feasibility phase. Uncontrolled changes during this phase would lead to scope creep.

R3-109 Which of the following resources has the **GREATEST** risk of failure while implementing any security solution?

A. Security hardware
B. Security staff
C. Security processes
D. Security software

B is the correct answer.

Justification:

A. Security hardware failure is a risk to the enterprise, but not the greatest risk. The misconfiguration of the hardware by staff is the greatest risk.
B. The human component has the greatest risk of failure because people are vulnerable to risk such as fraud and deliberate or accidental misconfiguration of software processes or hardware.
C. Security process failures are a risk to the enterprise, but failure to follow the process by staff is the greatest risk.
D. Security software failures are a risk to the enterprise, but most failures come from misuse or misconfiguration of the software, not from flaws in the software.

R3-110 Due to changes in the IT environment, the disaster recovery plan of a large enterprise has been modified. What is the **GREATEST** benefit of testing the new plan?

A. To ensure that the plan is complete
B. To ensure that the team is trained
C. To ensure that all assets have been identified
D. To ensure that the risk assessment was validated

A is the correct answer.

Justification:
A. The greatest benefit of testing the new plan is to ensure that the plan is complete and will work during a crisis. Testing ensures that all assets in scope were incorporated into the plan, that all staff are trained and familiar with their roles, and the backups have been tested.
B. While training the team is a benefit of testing, it is not the greatest benefit.
C. While ensuring that all assets have been identified is a benefit of testing, it is not the greatest benefit.
D. Testing can be a method of validating the risk assessment. However, the greatest benefit of testing the new plan is to ensure that the plan is complete and viable in the event of a crisis.

R3-111 The database administrator has decided to disable certain normalization controls in the database to provide users with increased query performance. This will **MOST** likely increase the risk of:

A. loss of audit trails.
B. duplicate indexes.
C. data redundancy.
D. unauthorized access to data.

C is the correct answer.

Justification:
A. Disabling normalization controls will have no impact on the audit trail.
B. Normalization will not address the creation of duplicate database indexes.
C. Normalization is performed to reduce redundant data; if these normalization controls are disabled to increase query performance, it will also increase the risk of redundancy of data.
D. Disabling normalization should not result in unauthorized access to data.

R3-112 What control focuses directly on preventing the risk of collusion?

A. Mandatory access control
B. Principle of least privilege
C. Discretionary access control
D. Mandatory job rotation

D is the correct answer.

Justification:
A. Mandatory access control is mandated by security policy. This will not prevent the risk of collusion.
B. Even if the principle of least privilege is followed, it will not be adequate to prevent collusion risk because two or more parties come together in collusion and have a higher privilege.
C. Discretionary access control is access as provided by the data owner and does not mean that two people within the same department cannot bypass segregation of duties.
D. Collusion risk happens when two or more people who are bound by segregation of duties come together and work together to bypass security controls. In job rotation, people will be assigned different job responsibilities over a period of time and this will reduce the opportunity for collusion.

R3-113 What is the purpose of system accreditation?

A. To ensure that risk associated with implementation has been identified and explicitly accepted by a senior manager
B. To review all technical and nontechnical controls to ensure that the security risk has been reduced to acceptable levels
C. To ensure that changes to the security controls are properly authorized, tested and documented
D. To require the training and certification of staff that will be responsible for working on a system

A is the correct answer.

Justification:

A. Accreditation is the decision to implement a system by an authorized senior manager based on the acceptance of risk as determined by the requirements.
B. Reviewing all technical and nontechnical controls to ensure that the security risk has been reduced to acceptable levels is the definition of certification. Certification is done to provide information to senior management that will be needed to make the accreditation decision.
C. Ensuring that changes to the security controls are properly authorized, tested and documented is the definition of change control. Change control will prevent unauthorized changes that may invalidate the accreditation.
D. Requiring the training and certification of staff is a responsibility related to implementing and maintaining an information system, but is not the purpose of accreditation.

R3-114 How does an enterprise **BEST** ensure that developers do not have access to implement changes to production applications?

A. The enterprise must ensure that development staff does not have access to executable code.
B. The enterprise must have segregation of duties between application development and operations.
C. The enterprise system development life cycle (SDLC) must be enforced to require segregation of duties.
D. The enterprise's change management process must be enforced for all but emergency changes.

B is the correct answer.

Justification:

A. The development staff should not have access to production systems. This is best managed through segregation of duties.
B. Segregation of duties will ensure that the developer cannot move a change into production. The developer can make the change, but the operations staff will only move the changed code into production upon approval through the change control process.
C. The system development life cycle (SDLC) is not used to implement segregation of duties, and will not prevent unauthorized change.
D. There must be a change control process for all changes, even emergency changes, otherwise the developer could move an unauthorized change into production under the guise of an emergency fix.

R3-115 How can an enterprise prevent duplicate processing of a transaction?

A. By encrypting the transaction to prevent copying
B. By comparing hash values of each transaction
C. By not allowing two identical transactions within a set time period
D. By not allowing more than one transaction per account per login

C is the correct answer.

Justification:
A. Encrypting a transaction will not prevent it from being processed twice.
B. An enterprise will often keep a hash of previous files, but not of individual transactions.
C. Any time that more than one identical transaction attempts to execute within a set time period, the second transaction should trigger a notification or a fraud alert.
D. Many services allow more than one transaction per login session.

R3-116 The **PRIMARY** goal of certifying a system prior to implementation is to:

A. protect the enterprise from liability for releasing a substandard system.
B. review the system controls to ensure that the controls are configured correctly.
C. test the integrated system to detect any upstream or downstream liabilities.
D. ensure that the system meets its specified security requirements at the time of testing.

D is the correct answer.

Justification:
A. While protecting the enterprise from liability is important, this is not the primary goal of the certification and accreditation process.
B. The configuration of the controls has not yet been tested because the system has not been implemented, and they must be tested prior to certification.
C. Testing the integrated system is an important part of the testing of the system, but that is not the primary goal of the process.
D. According to the certification and accreditation process, the goal is to deliver a system that meets the agreed-on set of security requirements and the operational conditions that were set for its implementation to ensure that it will be operated in a secure manner.

R3-117 Which of the following is a **PRIMARY** role of the system owner during the accreditation process? The system owner:

A. reviews and approves the security plan supporting the system.
B. selects and documents the security controls for the system.
C. assesses the security controls in accordance with the assessment procedures.
D. determines whether the risk to the business is acceptable.

B is the correct answer.

Justification:
A. The review and approval of the security plan, including system and general IT controls, is the responsibility of senior management, or a delegated authorized individual, not the system owner.
B. The system owner specifies the information security controls for the system being deployed based on functional requirements from the information owner.
C. The system owner does not test the controls. Security control testing is the responsibility of the security control assessor or an otherwise independent party.
D. Senior management is accountable for determining whether the risk to the business is acceptable.

R3-118 Which of the following compensating controls should management implement when a segregation of duties conflict exists because an enterprise has a small IT department?

A. Independent analysis of IT incidents
B. Entitlement reviews
C. Independent review of audit logs
D. Tighter controls over user provisioning

C is the correct answer.

Justification:

A. Independent analysis of IT incidents could point to segregation of duties violations. This is not a compensating control, but a detective control.
B. Entitlement reviews are performed to review the access of individuals to ensure that they have the proper access for their current role. This review is the responsibility of the data owner and usually occurs at regular intervals. This is not the best way to prevent or detect a segregation of duties conflict.
C. An independent review of the audit logs would be the best compensating control because someone outside the IT department is validating that actual activity did not exploit segregation of duties.
D. User provisioning is the process of granting access to an application or system. While a normal part of the provisioning process is to make sure that no segregation of duties conflicts exist, this cannot be done in the present case due to the small size of the IT department; therefore, tighter controls over user provisioning will be of limited value.

R3-119 The **BEST** way to ensure that an information systems control is appropriate and effective is to verify:

A. that the control is operating as designed.
B. that the risk associated with the control is being mitigated.
C. that the control has not been bypassed.
D. the frequency at which the control logs are reviewed.

B is the correct answer.

Justification:

A. A control may be operating correctly, but may not mitigate the risk it was designed to address. It is most important that the control reduces the risk it was designed to mitigate.
B. A control is designed to mitigate or reduce a risk. Even if the control is operating correctly, it is not the correct control if it does not address the risk it was designed to mitigate.
C. Even if the control has not been bypassed, it still may not effectively mitigate the associated risk.
D. A control must be checked periodically, but this does not ensure that it is the correct control to mitigate the risk.

R3-120 Which of the following information systems controls is the **BEST** way to detect malware?

A. Reviewing changes to file size
B. Reviewing administrative-level changes
C. Reviewing audit logs
D. Reviewing incident logs

A is the correct answer.

Justification:

A. One method to detect malware is to compare current executables and files with historical sizes and time stamps.
B. Administrative-level changes will not detect the presence of malware. They will provide a trigger to investigate depending on the number of administrative-level changes.
C. Audit logs do not hold data at a granular enough level to enable malware discovery.
D. Incident logs are used to identify a root cause that contributed to the introduction of malware.

R3-121 An enterprise security policy is an example of which control?

A. Operational control
B. Management control
C. Technical control
D. Corrective control

B is the correct answer.

Justification:

A. Manufacturing procedures are generally categorized as operational controls.
B. There are two control methods: technical and nontechnical. Enterprise security policies are nontechnical management controls.
C. Encryption and intrusion detection systems (IDSs) are typical examples of technical controls.
D. Corrective controls are used to respond to an incident and are not an example of policy.

R3-122 What is the **BEST** action to take once a new control has been implemented to mitigate a previously identified risk?

A. Update the risk register to show that the risk has been mitigated.
B. Schedule a new risk review to ensure that no new risk is present.
C. Test the control to ensure that the risk has been adequately mitigated.
D. Validate the tests conducted by the implementation team and close out the risk.

C is the correct answer.

Justification:

A. The risk assessment team cannot just accept assurance from another group without validating that the control did in fact adequately test and mitigate the risk.
B. There is no need to schedule a full, new risk review, but the remediated control should be tested.
C. The risk assessment team is responsible for the risk identification and cannot accept assurances from others that the risk has been adequately addressed. They must test to ensure the risk has in fact been properly mitigated.
D. Line items within the risk register can only be closed once the risk assessment team is assured of the effectiveness of the controls. The risk assessment team must run their own tests, not just validate test results conducted by the implementation team.

R3-123 What is the **BEST** tool for documenting the status of risk mitigation and risk ownership?

A. Risk action plans
B. Risk scenarios
C. Business impact analysis (BIA) documents
D. A risk register

D is the correct answer.

Justification:
A. Risk action plans define risk activities for a defined scope, not for an entire entity.
B. Risk scenarios help develop a thorough understanding of an enterprise's risk profile; however, they are not suitable for capturing the risk mitigation, contingency plans and ownership for an enterprise.
C. The business impact analysis (BIA) documents show ownership and describe processes and assets that are critical to the business; they do not describe risk mitigation strategies or specifically lay out the technical details of the contingency plan.
D. A risk register is designed to document all risk identified for the enterprise. For each risk it records, at a minimum: likelihood, potential impact, priority, status of mitigation and owner.

R3-124 Which of the following would data owners be **PRIMARILY** responsible for?

A. Intrusion detection
B. Antivirus controls
C. User entitlement changes
D. Platform security

C is the correct answer.

Justification:
A. Data custodians are responsible for designing and implementing intrusion detection, based on business needs.
B. Data custodians are responsible for designing and implementing antivirus controls, based on business needs.
C. Data owners are responsible for assigning user entitlement changes and approving access to the systems for which they are responsible.
D. Data custodians are responsible for designing and implementing platform security, based on business needs.

R3-125 When using a formal approach to respond to a security-related incident, which of the following provides the **GREATEST** benefit from a legal perspective?

A. Proving adherence to statutory audit requirements
B. Proving adherence to corporate data protection requirements
C. Demonstrating due care
D. Working with law enforcement agencies

C is the correct answer.

Justification:

A. There are several reasons why adherence to statutory audit requirements will not provide the greatest benefit:
 • Statutory audit requirements do not dictate information security or IT policies, which define the controls in place to protect the enterprise from security incidents.
 • The response to audit requirements is often a reactive approach.

B. One must implement security to ensure data protection. While this is a good option, it will not offer the greatest benefit because enterprises can have data protection requirements (i.e., policies, etc.) but fail to implement the information security actions to ensure data protection.

C. In the field of information security, the following statements are useful: *"Due care are steps that are taken to show that a company has taken responsibility for the activities that take place within the corporation and has taken the necessary steps to help protect the company, its resources, and employees."* And, *"continual activities that make sure the protection mechanisms are continually maintained and operational."* (Source: Harris, Shon; *All-in-one CISSP Certification Exam Guide, 2nd Edition*, McGraw-Hill/Osborne, USA, 2003.) Stockholders, customers, business partners and governments have the expectation that corporate officers will run the business in accordance with accepted business practices and in compliance with laws and other regulatory requirements. So while no entity can protect themselves completely from security incidents, in case of legal action, by demonstrating due care, these entities can make a case that they are actually doing things to monitor and maintain the protection mechanisms and that these activities are ongoing.

D. Working with law enforcement agencies often occurs after a breach or security incident happens, i.e., a reactive approach. While it is a commendable action, it still raises the question, often in the court of law, of whether the enterprise did everything it could to prevent the breach or incident.

R3-126 Which of the following processes is **CRITICAL** for deciding prioritization of actions in a business continuity plan (BCP)?

A. Risk assessment
B. Vulnerability assessment
C. A business impact analysis (BIA)
D. Business process mapping

C is the correct answer.

Justification:

A. Risk assessment provides information on the likelihood of occurrence of security incidents and assists in the selection of countermeasures, but not in the prioritization of actions.

B. A vulnerability assessment provides information regarding the security weaknesses of the system and supporting the risk analysis process.

C. The business impact analysis (BIA) is the most critical process for deciding which part of the information system/business process should be given prioritization in case of a security incident.

D. Business process mapping does not help in making a decision, but in implementing a decision.

R3-127 The **MOST** effective starting point to determine whether an IT system continues to meet the enterprise's business objectives is to conduct interviews with:

A. executive management.
B. IT management.
C. business process owners.
D. external auditors.

C is the correct answer.

Justification:
A. Executive management will be able to provide the overall picture of the enterprise's business objectives.
B. IT management is important, but should not be the starting point because they likely do not see a clear picture of all organizational objectives or how the business plans to use IT in the future.
C. Business process owners are an effective starting point for conducting interviews to ensure that IT systems are meeting their individual business process needs.
D. External auditors can be useful for an objective view on control performance of the IT systems, but they are not a starting point in determining if an IT system continues to meet organizational objectives.

R3-128 It is **MOST** important for risk mitigation to:

A. eliminate threats and vulnerabilities.
B. reduce the likelihood of risk occurrence.
C. reduce risk within acceptable cost.
D. reduce inherent risk to zero.

C is the correct answer.

Justification:
A. Threats are often outside the reach of the enterprise's influence and cannot be affected. Vulnerabilities can be reduced, but cannot be eliminated.
B. The likelihood of risk occurrence depends on many factors, many of which cannot be influenced internally.
C. Risk should be reduced or mitigated at an acceptable cost while reducing risk to an acceptable level.
D. Inherent risk of any activity cannot be affected. It is the risk level or the exposure without taking into account the actions that management has taken or might take (e.g., implementing controls).

R3-129 The cost of mitigating a risk should not exceed the:

A. expected benefit to be derived.
B. annual loss expectancy (ALE).
C. value of the physical asset.
D. cost to exploit the weakness.

A is the correct answer.

Justification:

A. **The cost of mitigating a risk should never exceed the value that is expected to result from its implementation. It is illogical to spend US $1,000 to protect against a risk that in a worst case scenario would create a loss of less than US $100.**

B. Annual loss expectancy (ALE) is incorrect because the remoteness of the likelihood may cause the ALE to be quite low. However, it may be worthwhile to spend an amount in excess of the ALE to protect against a loss that, if it occurred, would be significantly higher.

C. It may be worthwhile to spend more than the value of a physical asset when that asset contains something of even higher value. The value of a backup tape is not so much the cost of the tape as it is the value of what is stored on that tape.

D. The cost to exploit a weakness may be very low compared to its impact. For example, a freely available exploit from the Internet can be used to execute a denial-of-service attack on an e-commerce site. The amount that an enterprise spends on risk mitigation must be directly related to the likelihood and impact of a specific risk and how the control mitigates that risk.

R3-130 During the initial phase of the system development life cycle (SDLC), the risk professional provided input on how to secure the proposed system. The project team prepared a list of requirements that will be used to design the system. Which of the following tasks **MUST** be performed before moving on to the system design phase?

A. The risk associated with the proposed system and controls is accepted by management.
B. Various test scenarios that will be used to test the controls are documented.
C. The project budget is increased to include additional costs for security.
D. Equipment and software are procured to meet the security requirements.

A is the correct answer.

Justification:

A. **The risk acceptance decision is made by senior management. Before moving further into the project, it is important to have sign-off from management that management acknowledges and accepts the risk that is associated with this project. If management does not accept the risk, then there is no point in proceeding any further.**

B. As risk is being identified, it is good to begin developing scenarios to test the system against that risk, but this is not a critical step before moving into the design phase.

C. At the end of each phase, a go/no go decision should be made by management based on project feasibility and risk. However, it may not be necessary to revise the budget at this time.

D. It is too early in the process to begin the procurement of system components.

R3-131 Which of the following risk assessment outputs is **MOST** suitable to help justify an organizational information security program?

A. An inventory of risk that may impact the enterprise
B. Documented threats to the enterprise
C. Evaluation of the consequences
D. A list of appropriate controls for addressing risk

D is the correct answer.

Justification:

A. A risk inventory is not the best choice because it does not sufficiently address how the risk will be addressed.
B. Documentation of threats is not the best choice because it does not sufficiently address how the threats may exploit vulnerabilities and how the resulting risk will be reduced.
C. Evaluation of the consequences of a risk—in combination with the likelihood of a risk—is important for the prioritization of risk responses. However, it is not the best choice because it does not sufficiently address how the risk will be addressed.
D. A list of appropriate information security controls in response to the risk scenarios identified during the risk assessment is one of the primary deliverables of a risk assessment exercise. In this case it is also the best choice because it demonstrates due consideration of the risk as well as suitable controls to address the risk.

R3-132 Which of the following is the **MOST** desirable strategy when developing risk mitigation options associated with the unavailability of IT services due to a natural disaster?

A. Assume the worst-case incident scenarios.
B. Target low-cost locations for alternate sites.
C. Develop awareness focused on natural disasters.
D. Enact multiple tiers of authority delegation.

A is the correct answer.

Justification:

A. To be prepared for a natural disaster, it is appropriate to assume the worst-case scenario; otherwise, the resulting impact may exceed the enterprise's ability to recover.
B. Setting up a low-cost location for an alternate site may not always be a good strategy against natural disasters. Adequate investment should be made based on an impact analysis.
C. An awareness training program is a key factor for business continuity. However, its effectiveness may be limited.
D. Delegation of authority will work somewhat in case of emergency. However, this may be a situational decision in the event of natural disaster.

R3-133 What is the **PRIMARY** objective of conducting a peer review prior to implementing any changes to the firewall configuration?

A. To assist in the detection of fraudulent or inappropriate activity
B. To reduce the need for more technical testing since the changes have already been examined
C. To facilitate ongoing knowledgeable transfer staff to learn by examining the work of senior staff
D. To help detect errors in the proposed change prior to implementation

D is the correct answer.

Justification:

A. Peer review can help detect fraud or inappropriate activity, but this would rarely apply to a change to a firewall configuration.
B. Peer review is only one part of the change control process and does not remove the requirement to thoroughly test the change before and following implementation.
C. Peer review is the examination of a work product by a skilled expert and is not used for training new staff.
D. Peer review is the examination of a work product by a skilled coworker. This should highlight any errors or cases where standards are not being followed and may prevent the introduction of an error into production.

R3-134 The aggregated results of continuous monitoring activities are **BEST** communicated to:

A. the risk owner.
B. technical staff.
C. the audit department.
D. the information security manager.

A is the correct answer.

Justification:

A. The risk owner is the most suitable target audience for aggregated results of continuous monitoring because they own the risk and are accountable for the fact that appropriate risk responses are executed in alignment with the enterprise's risk appetite.
B. Technical staff are more likely to benefit from granular information from continuous monitoring that will enable them to address specific technical issues.
C. The audit department may review the continuous monitoring process as part of their audit activities, but is generally not the recipient of continuous monitoring reports.
D. The information security manager will receive the aggregated results for continuous monitoring for informational purposes, but is not the primary audience.

R3-135 An operations manager assigns monitoring responsibility of key risk indicators (KRIs) to line staff. Which of the following is **MOST** effective in validating the effort?

A. Reported results should be independently reviewed.
B. Line staff should complete risk management training.
C. The threshold should be determined by risk management.
D. Indicators should have benefits that exceed their costs.

A is the correct answer.

Justification:

A. **Because key risk indicators (KRIs) are monitored by line staff, there is a chance that staff may alter results to suppress unfavorable results. Additional reliability of monitoring metrics can be achieved by having the results reviewed by an independent party.**
B. It is not mandatory that line staff complete risk management training in order to be engaged in monitoring of KRIs.
C. The threshold should be determined through discussion between risk management and line staff/business managers.
D. It is important that the benefits of KRIs justify their costs; however, this determination does not help verify that the monitoring efforts of KRIs are effective.

DOMAIN 4—RISK AND CONTROL MONITORING AND REPORTING (22%)

R4-1 Which of the following is the **MOST** important reason for conducting periodic risk assessments?

A. Risk assessments are not always precise.
B. Reviewers can optimize and reduce the cost of controls.
C. Periodic risk assessments demonstrate the value of the risk management function to senior management.
D. Business risk is subject to frequent change.

D is the correct answer.

Justification:
A. Although an assessment can never be perfect and invariably contains some errors, this is not the most important reason for periodic reassessment.
B. Optimizing control cost is an insufficient reason.
C. Demonstrating the value of the risk management function to senior management is an insufficient reason.
D. Risk is constantly changing, so a previously conducted risk assessment may not include measured risk that has been introduced since the last assessment.

R4-2 Which of the following is **MOST** essential for a risk management program to be effective?

A. New risk detection
B. A sound risk baseline
C. Accurate risk reporting
D. A flexible security budget

A is the correct answer.

Justification:
A. Without identifying new risk, other procedures will only be useful for a limited period.
B. A risk baseline is essential for implementing risk management, but new risk detection is the most essential.
C. Accurate risk reporting is essential for implementing risk management, but new risk detection is the most essential.
D. A flexible security budget is not a reality for most enterprises. A limited security budget is a scope limitation that an effective risk management program works with by prioritizing risk responses.

R4-3 A network vulnerability assessment is intended to identify:

A. security design flaws.
B. zero-day vulnerabilities.
C. misconfigurations and missing updates.
D. malicious software and spyware.

C is the correct answer.

Justification:
A. Security design flaws require a deeper level of analysis.
B. Zero-day vulnerabilities, by definition, are not previously known and, therefore, are undetectable.
C. A network vulnerability assessment intends to identify known vulnerabilities that are based on common misconfigurations and missing updates.
D. Malicious software and spyware are normally addressed through antivirus and antispyware policies.

R4-4 Previously accepted risk should be:

A. reassessed periodically because the risk can be escalated to an unacceptable level due to revised conditions.
B. removed from the risk log once it is accepted.
C. accepted permanently because management has already spent resources (time and labor) to conclude that the risk level is acceptable.
D. avoided next time because risk avoidance provides the best protection to the enterprise.

A is the correct answer.

Justification:

A. Accepted risk should be reviewed regularly to ensure that the initial risk acceptance rationale is still valid within the current business context.
B. Even risk that has been accepted should be monitored for changing conditions that could alter the original decision.
C. The rationale for the initial risk acceptance may no longer be valid due to change(s), and therefore, risk cannot be accepted permanently.
D. Risk is an inherent part of business, and it is impractical and costly to eliminate all risk.

R4-5 After a risk assessment study, a bank with global operations decided to continue doing business in certain regions of the world where identity theft is widespread. To **MOST** effectively deal with the risk, the business should:

A. implement monitoring techniques to detect and react to potential fraud.
B. make the customer liable for losses if the customer fails to follow the bank's advice.
C. increase its customer awareness efforts in those regions.
D. outsource credit card processing to a third party.

A is the correct answer.

Justification:

A. Implementing monitoring techniques that will detect and deal with potential fraud cases is the most effective way to deal with this risk.
B. While making the customer liable for losses is a possible approach, the bank needs to be seen as proactive in managing its risk.
C. While customer awareness will help mitigate the risk, this is not sufficient on its own to control fraud risk.
D. If the bank outsources its processing, the bank still retains liability.

R4-6 Which of the following **BEST** indicates a successful risk management practice?

A. Control risk is tied to business units.
B. Overall risk is quantified.
C. Residual risk is minimized.
D. Inherent risk is eliminated.

C is the correct answer.

Justification:

A. Although tying control risk to business units may improve accountability, it is not as desirable as minimizing residual risk.
B. The fact that overall risk has been quantified does not necessarily indicate the existence of a successful risk management practice.
C. A successful risk management practice minimizes the residual risk to the enterprise.
D. It is virtually impossible to eliminate inherent risk.

R4-7 Which of the following **MOST** enables risk-aware business decisions?

A. Robust information security policies
B. An exchange of accurate and timely information
C. Skilled risk management personnel
D. Effective process controls

B is the correct answer.

Justification:

A. Security policies generally focus on protecting the business and do not enable risk-aware business decisions, particularly when the decision affects future business needs.
B. An exchange of information is a key area for management to be able to make risk-related decisions. Accuracy and timeliness of information are success factors.
C. Skilled risk management personnel enable risk-aware business decisions, but ideally the flow of information needs to be two-directional, ensuring that risk, loss and vulnerability events are reported and allowing risk management personnel to understand changes in the organization's risk appetite and tolerance.
D. Process controls generally exist for known threats and do not enable risk-based business decisions. Control monitoring, however, involves the dissemination of control information to enable a timely risk response (business decision).

R4-8 Which of the following should be of **MOST** concern to a risk practitioner?

A. Failure to notify the public of an intrusion
B. Failure to notify the police of an attempted intrusion
C. Failure to internally report a successful attack
D. Failure to examine access rights periodically

C is the correct answer.

Justification:

A. Reporting to the public is not a requirement and is dependent on the enterprise's desire, or lack thereof, to make the intrusion known.
B. It is highly unlikely that an attempted intrusion requires notification of the police. Moreover, attempted intrusions are not as significant to the risk practitioner as activities related to successful attacks.
C. Failure to report a successful intrusion is a serious concern to the risk practitioner and could—in some instances—be interpreted as abetting.
D. Although the lack of a periodic examination of access rights may be a concern, it does not represent as big a concern as the failure to report a successful attack.

R4-9 Which of the following is the **FIRST** step when developing a risk monitoring program?

A. Developing key indicators to monitor outcomes
B. Gathering baseline data on indicators
C. Analyzing and reporting findings
D. Conducting a capability assessment

D is the correct answer.

Justification:

A. Developing key indicators to monitor outcomes is necessary, but not the first step. There is no use for indicators if there is no information on what these indicators are going to report.
B. Gathering baseline data on indicators is necessary, but not the first step. There is no use for gathering baseline data if the indicators are not defined.
C. Analyzing and reporting findings is necessary, but not the first step. There is no use for analyzing and reporting findings if the baseline is not there.
D. This step determines the capacity and readiness of the entity to develop a risk management program. This assessment identifies champions, barriers, owners and contributors to this program, including identifying the overall goal of the program. A capability assessment helps determine the enterprise's maturity in its risk management processes and the capacity and readiness of the entity to develop a risk management program. When the enterprise is more mature, more sophisticated responses can be implemented; when the enterprise is rather immature, some basic responses may be a better starting point.

R4-10 Which of the following reviews will provide the **MOST** insight into an enterprise's risk management capabilities?

A. A capability maturity model (CMM) review
B. A capability comparison with industry standards or regulations
C. A self-assessment of capabilities
D. An internal audit review of capabilities

A is the correct answer.

Justification:

A. Capability maturity modeling allows an enterprise to understand its level of maturity in its risk capabilities, which is an indicator of operational readiness and effectiveness.
B. A capability comparison with industry standards or regulations does not provide insights into readiness and effectiveness, but only into the existence or nonexistence of capabilities exclusive of maturity.
C. A self-assessment of capabilities does not provide insights into readiness and effectiveness, but only into the existence or nonexistence of capabilities exclusive of maturity.
D. An internal audit review of capabilities does not provide insights into readiness and effectiveness, but only into the existence or nonexistence of capabilities exclusive of maturity.

R4-11 Which of the following practices is **MOST** closely associated with risk monitoring?

A. Assessment
B. Mitigation
C. Analysis
D. Reporting

D is the correct answer.

Justification:
A. Risk assessment is associated with risk identification and evaluation, but not with risk monitoring.
B. Risk mitigation is associated with risk response, but not with risk monitoring.
C. Risk analysis is associated with risk identification and evaluation, but not with risk monitoring.
D. Risk reporting is the only activity listed that is typically associated with risk monitoring.

R4-12 As part of an enterprise risk management (ERM) program, a risk practitioner **BEST** leverages the work performed by an internal audit function by having it:

A. design, implement and maintain the ERM process.
B. manage and assess the overall risk awareness.
C. evaluate ongoing changes to organizational risk factors.
D. assist in monitoring, evaluating, examining and reporting on controls.

D is the correct answer.

Justification:
A. The design, implementation and maintenance of the enterprise risk management (ERM) function is the responsibility of management, not of the internal audit function.
B. Overall risk awareness is the responsibility of the risk governance function.
C. Evaluating ongoing changes to the enterprise is not the responsibility of the internal audit function.
D. The internal audit function is responsible for assisting management and the board of directors in monitoring, evaluating, examining and reporting on internal controls, regardless of whether an ERM function has been implemented.

R4-13 Where are key risk indicators (KRIs) **MOST** likely identified when initiating risk management across a range of projects?

A. Risk governance
B. Risk response
C. Risk analysis
D. Risk monitoring

B is the correct answer.

Justification:

A. Risk governance is a systemic approach to decision-making processes associated with risk. From a CRISC perspective, information technology risk is adopted to achieve more effective risk management and to reduce risk exposure and vulnerability by filling gaps in the risk policy. This is not the best answer because it is not a risk management activity, but rather a risk management oversight function.

B. Key risk indicators (KRIs) and risk definition and prioritization are both considered part of the risk response process. After having identified, quantified and prioritized the risk to the enterprise, relevant risk indicators need to be identified to help provide risk owners with meaningful information about a specific risk, or a combination of types of risk.

C. Risk analysis is the process of identifying the types, probability and severity of risk that may occur during a project. Once the identification has taken place, the analysis feeds into the risk response process where one of the tasks is to identify KRIs.

D. Risk monitoring occurs after the risk response process and is ongoing. Assigning ownership to KRIs and defining various levels of KRI thresholds—along with automating the monitoring and notification process—help ensure monitoring of KRIs. KRIs must be identified before risk monitoring is implemented.

R4-14 Which of the following can be expected when a key control is being maintained at an optimal level?

A. The shortest lead time until the control breach comes to the surface
B. Balance between control effectiveness and cost
C. An adequate maturity level of the risk management process
D. An accurate estimation of operational risk amounts

B is the correct answer.

Justification:

A. Even though a key control is in place, it may take time until a breach surfaces if escalation procedures are not adequately set up. Thus, a key control alone does not assure the shortest lead time for a breach to be communicated to management.

B. Maintaining controls at an optimal level translates into a balance between control cost and derived benefit.

C. Measurement of the maturity level in risk management may depend on the function of key controls. However, the key control is not the major driver to assess the maturity of risk management.

D. The key control does not directly contribute to the accurate estimation of operational risk amounts. Maintenance of an incident database and the application of statistical method are essential for the estimation of operational risk.

R4-15 The **PRIMARY** reason to report significant changes in IT risk to management is to:

A. update the information asset inventory on a periodic basis.
B. update the values of probability and impact for the related risk.
C. reconsider the degree of importance of existing information assets.
D. initiate a risk impact analysis to determine if additional response is required.

D is the correct answer.

Justification:
A. This choice is not correct because an asset inventory may be updated even when there is no significant risk reported and because this is less important than initiating an appropriate risk response for impacted information assets.
B. This choice is not correct because updating new probability and impact values may happen when the risk assessment is performed or when significant risk is identified and analyzed.
C. This choice is not correct because management staff of relevant functions understand the importance of their assets and do not wait for significant risk to reconsider this fact.
D. The changes in information risk will impact the business process of a department or multiple departments and the security manager should report this to department heads so that they are able to initiate a risk analysis to determine the impact and if there are changes needed.

R4-16 A database administrator notices that the externally hosted, web-based corporate address book application requires users to authenticate, but that the traffic between the application and users is not encrypted. The **MOST** appropriate course of action is to:

A. notify the business owner and the security manager of the discovery and propose an addition to the risk register.
B. contact the application administrators and request that they enable encryption of the application's web traffic.
C. alert all staff about the vulnerability and advise them not to log on from public networks.
D. accept that current controls are suitable for nonsensitive business data.

A is the correct answer.

Justification:
A. The business owner and security manager should be notified and the risk should be documented on the operational or security risk register to enable appropriate risk treatment.
B. Enabling encryption without further assessment and input from the business owner is inappropriate because the vulnerability may indicate further issues with security that need to be resolved.
C. Alerting all of the staff without discussing the risk with the business owner and having a plan to rectify the issue can be damaging.
D. The database administrator is not the owner of the corporate address book application and therefore does not have authority to accept business risk.

R4-17 Which of the following is the **PRIMARY** reason for periodically monitoring key risk indicators (KRIs)?

A. The cost of risk response needs to be minimized.
B. Errors in results of KRIs need to be minimized.
C. The risk profile may have changed.
D. Risk assessment needs to be continually improved.

C is the correct answer.

Justification:
A. Minimizing the cost of risk response efforts can be one of the outcomes, but this is not the primary reason.
B. If there are errors in results of key risk indicators (KRIs), they can be minimized even without having periodic monitoring in place.
C. The current set of risk impacting the enterprise can change over time and periodic monitoring of KRIs proactively identifies changes in the risk profile so that new risk can be addressed and changes in levels in existing risk can be better controlled.
D. Risk assessment process improvements are not the reason for monitoring KRIs on a periodic basis.

R4-18 Which of the following is the **BEST** indicator of high maturity of an enterprise's IT risk management process?

A. People have appropriate awareness of risk and are comfortable talking about it.
B. Top management is prepared to invest more money in IT security.
C. Risk assessment is encouraged in all areas of IT and business management.
D. Business and IT are aligned in risk assessment and risk ranking.

A is the correct answer.

A. Some of the most important measures of a mature IT risk management process are those related to a risk-aware culture—an enterprise where people recognize the risk inherent to their activities, are able to discuss it and are willing to work together to resolve the risk.
B. While investment in IT security may strengthen the overall risk management posture of the enterprise, it is not an appropriate measure of IT risk management process maturity.
C. While risk assessment is an important step in the risk management process, it is not a good indicator of a mature risk management process, even when deployed across all business units and functions.
D. Alignment between IT and business is the foundation of an effective IT risk management process; however, it is not a good indicator of a mature IT risk management process.

R4-19 As part of risk monitoring, the administrator of a two-factor authentication system identifies a trusted independent source indicating that the algorithm used for generating keys has been compromised. The vendor of the authentication system has not provided further information. Which of the following is the **BEST** initial course of action?

A. Wait for the vendor to formally confirm the breach and provide a solution.
B. Determine and implement suitable compensating controls.
C. Identify all systems requiring two-factor authentication and notify their business owners.
D. Disable the system and rely on the single-factor authentication until further information is received.

C is the correct answer.

Justification:
A. Waiting on the vendor to acknowledge the vulnerability may result in an unacceptable exposure and may be considered negligent.
B. Determining suitable compensating controls is not appropriate without instructions from the responsible business owner.
C. Business owners should be notified, even when some of the information may not be available. Business owners are responsible for responding to new risk.
D. Disabling the system is not appropriate because there is no indication that the compromise will have an impact on the first-factor authentication.

R4-20 Which of the following is **MOST** useful in developing a series of recovery time objectives (RTOs)?

A. Regression analysis
B. Risk analysis
C. Gap analysis
D. Business impact analysis (BIA)

D is the correct answer.

Justification:
A. Regression analysis is used to test changes to program modules.
B. Risk analysis is the process by which frequency and impact of risk scenarios are estimated; it is a component of a business impact analysis (BIA).
C. Gap analysis is useful in addressing the differences between the current state and an ideal future state.
D. Recovery time objectives (RTOs) are a primary deliverable of a BIA. RTOs relate to the financial impact of a system not being available.

R4-21 Which of the following is the **BEST** way to ensure that contract programmers comply with organizational security policies?

A. Have the contractors acknowledge the security policies in writing.
B. Perform periodic security reviews of the contractors.
C. Explicitly refer to contractors in the security standards.
D. Create penalties for noncompliance in the contracting agreement.

B is the correct answer.

Justification:
A. Written acknowledgements of security policies do not help detect the failure of contract programmers to comply.
B. Periodic reviews are the most effective way of obtaining compliance.
C. Referring to the contract programs within security standards does not help detect the failure of contract programmers to comply with organizational security policies. It may establish responsibility for a control implementation and maintenance, but the control ownership and accountability remains within the enterprise itself.
D. Penalties do not help detect failure of contract programmers to comply with organizational security policies and can only be enforced once they are detected either by an audit or an incident.

R4-22 Management wants to ensure that IT is successful in delivering against business requirements. Which of the following **BEST** supports that effort?

A. An internal control system or framework
B. A cost-benefit analysis
C. A return on investment (ROI) analysis
D. A benchmark process

A is the correct answer.

Justification:
A. For IT to be successful in delivering against business requirements, management should develop an internal control system that supports its business requirements.
B. A cost-benefit analysis, although useful, is not the most important element to align IT to business.
C. A return on investment (ROI) analysis is just one of the metrics to measure success of IT investments.
D. A benchmark process comes into place once a sound internal control framework has been enabled.

R4-23 Which of the following is the **MOST** effective way to ensure that third-party providers comply with the enterprise's information security policy?

A. Security awareness training
B. Penetration testing
C. Service level monitoring
D. Periodic auditing

D is the correct answer.

Justification:
A. Training can increase user awareness of the information security policy, but is not more effective than auditing.
B. Penetration testing can identify security vulnerability, but cannot ensure information compliance.
C. Service level monitoring can only pinpoint operational issues in the enterprise's operational environment.
D. A regular audit exercise can spot any gap in the information security compliance.

R4-24 Which of the following metrics is the **MOST** useful in measuring the monitoring of violation logs?

A. Penetration attempts investigated
B. Violation log reports produced
C. Violation log entries
D. Frequency of corrective actions taken

A is the correct answer.

Justification:
A. The most useful metric is one that measures the degree to which complete follow-through has taken place.
B. Violation log reports are not indicative of whether investigative action was taken. The most useful metric is one that measures the degree to which complete follow-through has taken place.
C. Violation log entries are not indicative of whether investigative action was taken. The most useful metric is one that measures the degree to which complete follow-through has taken place.
D. Frequency of corrective actions taken is not indicative of whether investigative action was taken. The most useful metric is one that measures the degree to which complete follow-through has taken place.

R4-25 Which of the following is **MOST** important for measuring the effectiveness of a security awareness program?

A. Increased interest in focus groups on security issues
B. A reduced number of security violation reports
C. A quantitative evaluation to ensure user comprehension
D. An increased number of security violation reports

D is the correct answer.

Justification:
A. Focus groups may or may not provide meaningful feedback, but in and of themselves do not provide metrics.
B. A reduction in the number of violation reports may not be indicative of a high level of security awareness.
C. To judge the effectiveness of user awareness training, measurable testing is necessary to confirm user comprehension. However, comprehension of what needs to be done does not ensure that action is taken when necessary. The most effective indicator for measuring success of an awareness program is an increase in the number of violation reports by staff.
D. Of the choices offered, an increase in the number of violation reports is the best indicator of a high level of security awareness. As with automated alerts, each security violation report needs to be assessed for validity.

R4-26 When a significant vulnerability is discovered in the security of a critical web server, immediate notification should be made to the:

A. development team to remediate.
B. data owners to mitigate damage.
C. system owner to take corrective action.
D. incident response team to investigate.

C is the correct answer.

Justification:
A. The development team may be called on by the system owner to resolve the vulnerability.
B. Data owners are notified only if the vulnerability could have compromised data.
C. To correct the vulnerabilities, the system owner needs to be notified quickly, before an incident can take place.
D. Investigation by the incident response team is not correct because the incident has not taken place and notification could delay implementation of the fix.

R4-27 Which of the following is the **BEST** metric to manage the information security program?

A. The number of systems that are subject to intrusion detection
B. The amount of downtime caused by security incidents
C. The time lag between detection, reporting and acting on security incidents
D. The number of recorded exceptions from the minimum information security requirements

D is the correct answer.

Justification:
A. The number of systems subject to intrusion has no relevance to the quality of security management, but has more to do with the enterprise's vulnerability.
B. The amount of downtime is a measure of the scale of the threat.
C. The time lag is a measure of the responsiveness of the security team.
D. The number of exceptions from set requirements is a direct correlation to the quality of the security program.

R4-28 Which of the following **MOST** effectively ensures that service provider controls are within the guidelines set forth in the organization's information security policy?

A. Service level monitoring
B. Penetration testing
C. Security awareness training
D. Periodic auditing

D is the correct answer.

Justification:

A. Service level monitoring helps pinpoint the service provider's operational issues, but is not designed to ensure compliance.
B. Penetration testing helps identify system vulnerabilities, but is not designed to ensure compliance.
C. Security awareness training is a preventive measure to increase user awareness of the information security policy, but is not designed to ensure compliance.
D. Periodic audits help ensure compliance with the organization's information security policy.

R4-29 Despite a comprehensive security awareness program annually undertaken and assessed for all staff and contractors, an enterprise has experienced a breach through a spear phishing attack. What is the **MOST** effective way to improve security awareness?

A. Review the security awareness program and improve coverage of social engineering threats.
B. Launch a disciplinary process against the people who leaked the information.
C. Perform a periodic social engineering test against all staff and communicate summary results to the staff.
D. Implement a data loss prevention system that automatically points users to corporate policies.

C is the correct answer.

Justification:

A. The awareness program should be periodically reviewed, despite the incident. However, because the awareness program is comprehensive and undertaken by all staff, the review will not create the necessary improvement.
B. Spear phishing attacks are designed to look like justified communication from a trusted source; thus, a disciplinary process is inappropriate.
C. Users who are aware of security threats may need a reminder that these threats are real. Periodic social engineering tests help in maintaining a level of alertness.
D. Investment in the data loss prevention system may not be justified because it is designed to protect against a loss of special types of data and would not stop disclosure of user credentials.

R4-30 Which of the following is **MOST** critical when system configuration files for a critical enterprise application system are being reviewed?

A. Configuration files are frequently changed.
B. Changes to configuration files are recorded.
C. Access to configuration files is not restricted.
D. Configuration values do not impact system efficiency.

C is the correct answer.

Justification:

A. Changes to configuration files are based on business need and may be frequent. The key is validating that these changes are documented in the change management system and approved before being implemented into the production environment.
B. Even if the changes to the parameter file are recorded, this is less critical than access to configuration files because if access is not restricted, then the unauthorized user can disable recording of changes in the system using accounts with a high privilege.
C. If access to configuration files is not restricted, then the security of the overall system will be in question.
D. If access to the parameter file is not restricted, then the security of the overall system will be in question because access is bypassed; the system can be impacted in many ways and the efficiency of the system will be a lesser problem than losing control of the entire system.

R4-31 The **PRIMARY** reason for developing an enterprise security architecture is to:

A. align security strategies between the functional areas of an enterprise and external entities.
B. build a barrier between the IT systems of an enterprise and the outside world.
C. help with understanding of the enterprise's technologies and the interactions between them.
D. protect the enterprise from external threats and proactively monitor the corporate network.

A is the correct answer.

Justification:

A. The enterprise security architecture must align with the strategies and objectives of the enterprise, taking into consideration the importance of the free flow of information within an enterprise as well as business with partners, customers and suppliers.
B. Building a barrier between the IT systems of an enterprise and the outside world without proper alignment with business, information and technology may interfere with valid business processes.
C. The development of enterprise security architecture should not only take into consideration every piece of technology that exists in the enterprise, but also provide an understanding of how and why these technologies interact with each other as well as outside processes, suppliers, partners, customers and existing business processes to achieve enterprise objectives.
D. An enterprise security architecture does not protect the enterprise from threats nor does it perform monitoring of threats; it lays down a blueprint, including internal and external controls needed to protect the enterprise.

R4-32 During an organizational risk assessment it is noted that many corporate IT standards have not been updated. The **BEST** course of action is to:

A. review the standards against current requirements and make a determination of adequacy.
B. determine that the standards should be updated annually.
C. report that IT standards are adequate and do not need to be updated.
D. review the IT policy document and see how frequently IT standards should be updated.

A is the correct answer.

Justification:

A. **The risk practitioner should verify that the standards are still adequate. If standards are lacking, then they should be updated.**
B. Standards may or may not need to be updated, but should be reviewed annually for adequacy.
C. The risk practitioner cannot report that the IT standards are accurate until they are reviewed.
D. Reviewing the IT policy will not help determine whether the standards are still adequate or relevant.

R4-33 There is an increase in help desk call levels because the vendor hosting the human resources (HR) self-service portal has reduced the password expiration from 90 to 30 days. The corporate password policy requires password expiration after 60 days and HR is unaware of the change. The risk practitioner should **FIRST**:

A. formally investigate the cause of the unauthorized change.
B. request the service provider reverse the password expiration period to 90 days.
C. initiate a request to strengthen the corporate password expiration requirement to 30 days.
D. notify employees of the change in password expiration period.

A is the correct answer.

Justification:

A. **The key risk for the business process owner is that the external vendor is performing unauthorized changes to the configuration settings. All other actions are incorrect, because any change carries risk and requires a rigorous management approach.**
B. Reversing the change to a 90-day password expiration period would result in noncompliance with corporate policy.
C. While changing the corporate password policy to be more stringent may seem more secure, corporate policy should always be driven by business requirements, not the actions of service providers. Effective security requirements balance security with operational functionality.
D. While exceeding the requirements of the corporate password policy may seem more secure, such activities should follow a formal change management process and not be driven by individual actions or arbitrary changes from a service provider. Effective security requirements balance security with operational functionality.

R4-34 An excessive number of standard workstation images can be categorized as a key risk indicator (KRI) for:

A. change management.
B. configuration management.
C. IT operations management.
D. data management.

B is the correct answer.

Justification:
A. Change management deals with the process of managing changes to existing environments, rather than the initial environment definition.
B. An excessive number of unique workstation images is an indicator that poor configuration management processes are in place and that sufficient attention to actual business requirements has not been paid during the initial image definition.
C. IT operations management relates to the day-to-day operations of IT.
D. Data management relates to the handling of the data, rather than environment definition.

R4-35 Which of the following causes the **GREATEST** concern to a risk practitioner reviewing a corporate information security policy that is out of date? The policy:

A. was not reviewed within the last three years.
B. is missing newer technologies/platforms.
C. was not updated to account for new locations.
D. does not enforce control monitoring.

A is the correct answer.

Justification:
A. Not reviewing the policy for three years and updating it as necessary does not follow best practices and is the greatest concern.
B. Corporate information security policies are generally written at a level that does not require modification for specific, newer technologies and should not cause the greatest concern to the risk practitioner.
C. Corporate information security policies are generally written at a level that incorporates multiple locations, Even if the new facilities are in different geographic locations, with potentially different legislatures, a well-written corporate information security policy should accommodate such changes in the enterprise's operating environment.
D. Lack of control monitoring is a concern; however, the fact that the corporate information security policy itself was not reviewed on a regular basis is the greatest concern, particularly because policy reviews can be considered a part of continuous control monitoring at the highest level.

R4-36 Which of the following provides the **BEST** capability to identify whether controls that are in place remain effective in mitigating their intended risk?

A. A key performance indicator (KPI)
B. A risk assessment
C. A key risk indicator (KRI)
D. An audit

C is the correct answer.

Justification:

A. A key performance indicator (KPI) is a measure that determines how well the process enables the goal to be reached. A KPI is a lead indicator of whether a goal will likely be reached and a good indicator of capabilities, practices and skills. It measures an activity goal, which is an action that the process owner must take to achieve effective process performance.
B. A risk assessment is a process used to identify and evaluate risk and its potential effects. It includes assessing the critical functions necessary for an enterprise to continue business operations, defining the controls in place to reduce enterprise exposure and evaluating the cost for such controls. Risk analysis often involves an evaluation of the probabilities of a particular event.
C. A key risk indicator (KRI) identifies whether a risk exists and has the potential to be realized in such way that it will have a negative impact on the enterprise. If controls that are in place to mitigate identified risk are working properly, then KRIs should not report a concern.
D. An audit is a formal inspection and verification to check whether a standard or set of guidelines is being followed, records are accurate, or efficiency and effectiveness targets are being met.

R4-37 Which of the following is the **PRIMARY** reason for conducting periodic risk assessments?

A. Changes to the asset inventory
B. Changes to the threat and vulnerability profile
C. Changes in asset classification levels
D. Changes in the risk appetite

B is the correct answer.

Justification:

A. Changes in the asset inventory occur as assets move in and out of commission; they generally are not a trigger for a periodic risk assessment unless there are significant changes due to a system migration, data center build or other nonoperational event.
B. Changes in threats and vulnerabilities, including new occurrences of either, are the primary reasons to conduct periodic risk assessments.
C. Changes in asset classification levels are fairly rare and generally are not a trigger for periodic risk assessments. However, a low-value asset that is excluded from assessment may be reclassified as a regulatory-related or critical asset because it is supporting a high-value process; the reclassification would trigger a risk assessment and future periodic assessments for this asset.
D. Changes in risk appetite do not trigger a new risk assessment; however, they will affect how the business responds to risk.

R4-38 A risk practitioner has become aware of a potential merger with another enterprise. What action should the risk practitioner take?

A. Evaluate how the changes in the business operations and culture could affect the risk assessment.
B. Monitor the situation to see if any new risk emerges due to the proposed changes.
C. Continue to monitor and enforce the current risk program because it is already tailored appropriately for the enterprise.
D. Implement changes to the risk program to prepare for the transition.

A is the correct answer.

Justification:
A. Changes to the business may impact risk calculations, and the risk practitioner should be proactive and be prepared to deal with any changes as they happen.
B. The risk practitioner should continue to evaluate the risk levels, but should also evaluate how new risk may emerge as a result of changes to the business.
C. Risk assessment is a continuous process and should be revisited whenever a significant change is pending.
D. Because this is a potential change, the risk practitioner should prepare for, but not make, any changes to the risk program at this time.

R4-39 Which of the following **BEST** enables an enterprise to measure its risk management process against peers?

A. Adoption of an enterprise architecture (EA) model
B. Adoption of a balanced scorecard (BSC)
C. Adoption of a risk assessment methodology
D. Adoption of a maturity model

D is the correct answer.

Justification:
A. An enterprise architecture (EA) is unique to an enterprise.
B. A balanced scorecard (BSC) is unique to an enterprise.
C. Results of risk assessments will be enterprise-specific because no two business environments are the same.
D. A maturity model consists of various levels of competence that enterprises can use as benchmarks to assess how they compare to peers.

R4-40 A risk practitioner has collected several IT-related key risk indicators (KRIs) related for the core financial application. These would **MOST** likely be reported to:

A. stakeholders.
B. the IT administrator group.
C. the finance department.
D. senior management.

D is the correct answer.

Justification:

A. Stakeholders are a broad group of internal and external individuals and entities that are affected by a specific process. While some stakeholders may need to know about relevant key risk indicators (KRIs), it may not be suitable to share such information with other stakeholders.
B. The IT administrators group is not a key target for sharing IT-related KRIs. KRIs generally are shared with those who make risk response decisions or who are accountable for the execution of risk responses.
C. The finance department is not a key target for sharing IT-related KRIs for the financial application. KRIs generally are shared with those who make risk response decisions or who are accountable for the execution of risk responses.
D. Senior management is a key target group for sharing IT-related KRIs for the financial application because they make decisions related to risk response.

R4-41 An enterprise is expanding into new nearby domestic locations (office park). Which of the following is **MOST** important for a risk practitioner to report on?

A. Competitor analysis
B. Legal and regulatory requirements
C. Political issues
D. The potential of natural disasters

B is the correct answer.

Justification:

A. Competitors will most likely not change by moving from one domestic location to another nearby domestic location.
B. Legal and regulatory requirements are most likely to change when moving to a nearby location because each municipality may enforce significantly different regulations, including environmental requirements, taxation and others.
C. Expansion into new nearby domestic locations will most likely not cause exposure to political issues.
D. Expansion into new nearby domestic locations will most likely not cause exposure to new potential natural disasters.

R4-42 The **MOST** important reason for reporting control effectiveness as part of risk reporting is that it:

A. enables audit reporting.
B. affects the risk profile.
C. requires mitigation.
D. helps manage the control life cycle.

B is the correct answer.

Justification:
A. Changes in controls are not necessarily reported to the audit function.
B. Changes may render a control ineffective and allow a vulnerability to be exploited. Changes in control may also strengthen the enterprise's risk profile, e.g., in cases where highly manual process are automated.
C. Changes in controls may be that a weaker control is replaced by a stronger control; they may not necessarily require mitigation.
D. Reporting changes in controls may help manage the control life cycle, particularly in cases in which a control is failing and is consequently modified or replaced.

R4-43 Which of the following is **MOST** suitable for reporting IT-related business risk to senior management?

A. Balanced scorecards (BSCs)
B. Gantt charts/PERT diagrams
C. Technical vulnerability reports
D. Dashboards

D is the correct answer.

Justification:
A. A balanced scorecard (BSC) is a coherent set of performance measures organized into four categories that includes traditional financial measures, but adds customer, internal business process, and learning and growth perspectives.
B. Gantt charts/PERT charts show the critical path for a project, yet are not suitable for reporting IT-related business risk.
C. Technical vulnerability reports provide a detailed overview of system vulnerabilities and often leading practices on how to mitigate these vulnerabilities. Often, they are not tied to the business impact and are too granular to be used for reporting IT-related business risk to senior management.
D. Dashboards are most suitable for reporting risk to senior management because they provide a high-level overview of risk levels that can be easily understood.

R4-44 A key objective when monitoring information systems control effectiveness against the enterprise's external requirements is to:

A. design the applicable information security controls for external audits.
B. create the enterprise's information security policy provisions for third parties.
C. ensure that the enterprise's legal obligations have been satisfied.
D. identify those legal obligations that apply to the enterprise's security practices.

C is the correct answer.

Justification:
A. Control design occurs in the risk treatment phase instead of in the monitoring phase.
B. Creating the information security policy should occur well in advance of control monitoring.
C. Legal obligations are one of the principal external requirements to which compliance should be monitored.
D. The identification of the legal obligations should occur before risk treatment, so that the proper controls may be designed.

R4-45 Which of the following **BEST** helps the risk practitioner identify IS control deficiencies?

A. An IT control framework
B. Defined control objectives
C. A countermeasure analysis
D. A threat analysis

B is the correct answer.

Justification:

A. An IT control framework is generic and reviewing it does not help in identifying IS control deficiencies.
B. Controls are deployed to achieve the desired objectives based on risk assessment and to meet the business requirements.
C. A countermeasure analysis provides results on countermeasures for a control. The countermeasures are deployed when a threat is perceived and additional controls act as countermeasures. This, however, does not help to identify IS control deficiencies.
D. A threat analysis identifies the various threats affecting the systems and assets and does not help to identify IS control deficiencies.

R4-46 The **BEST** reason to implement a maturity model for risk management is to:

A. permit alignment with business objectives.
B. help improve governance and compliance.
C. ensure that security controls are effective.
D. enable continuous improvement.

D is the correct answer.

Justification:

A. Maturity models help benchmark processes and identify gaps between the current and the desired state of specific processes. They do not enable alignment with business objectives, which is more effectively achieved through a balanced scorecard or a goals cascade approach.
B. While maturity models help identify gaps between the current and the desired state of specific business processes, they do not explicitly improve governance and compliance efforts.
C. Maturity models help benchmark business processes and identify gaps between the current and the desired states. Maturity models to not explicitly ensure that security controls are effective.
D. Maturity models are designed to enable continuous improvement. This is achieved by first assessing the current maturity level of specific business processes and determining whether it is congruent with the desired maturity levels. Where gaps exist, maturity models implicitly provide steps to improve the process by defining requirements for each maturity level.

R4-47 Which of the following considerations is **MOST** important when implementing key risk indicators (KRIs)?

A. The metric is easy to measure.
B. The metric is easy to aggregate.
C. The metric is easy to interpret.
D. The metric links to a specific risk.

D is the correct answer.

Justification:
A. Ease of measuring the key risk indicator (KRI) is an important consideration and includes the consideration of data extraction, validation, aggregation and analysis. It is, however, secondary to linking a KRI to a specific risk.
B. An important consideration of metrics is the ability to classify and combine several metrics together in order to understand the underlying risk they represent. This is, however, secondary to linking a KRI to a specific risk.
C. Being able to easily understand (interpret) the metric is an important consideration. It is, however, secondary to linking a KRI to a specific risk.
D. Linking to a specific risk is the most important criterion when selecting a KRI.

R4-48 Which of the following data is **MOST** useful for communicating enterprise risk to management?

A. Control self-assessment results
B. A controls inventory
C. Key risk indicators (KRIs)
D. Independent audit reports

C is the correct answer.

Justification:
A. Creating economies of scale will allow for the enterprise to share common resources. This is typically done during the identification of business opportunities phase.
B. A controls inventory will assist the enterprise in managing risk more efficiently because existing controls can be considered during risk scenario development or when selecting a risk response.
C. Reporting on key risk indicators (KRIs) is the most useful for informing management of the current state of enterprise risk.
D. Independent audit reports provide insights on audit findings and related risk, based on the specific scope of the audits being performed. Audit reports do not provide an enterprisewide risk perspective.

R4-49 An enterprise has just completed an information systems audit and a large number of findings have been generated. This list of findings is **BEST** addressed by:

A. a risk mitigation plan.
B. a business impact analysis (BIA).
C. an incident management plan.
D. revisions to information security procedures.

A is the correct answer.

Justification:
A. This is the proper tool to address the identified risk. A risk mitigation plan will put forward a schedule and strategy for addressing the audit findings.
B. A business impact analysis (BIA) is a process to determine the impact of losing the support of any resource.
C. An incident management plan is used to prepare for, detect, respond to and mitigate the effects of incidents.
D. Revisions to information security procedures would likely only address a portion of the audit findings.

R4-50 What is the **PRIMARY** reason for reporting significant changes in information risk to senior management?

A. To revise the key risk indicators (KRIs)
B. To enable educated decision making
C. To gain support for new countermeasures
D. To recalculate the value of existing information assets

B is the correct answer.

Justification:
A. Revisions in key risk indicators (KRIs) have to be communicated to management; however, they are not the way to communicate significant new information risk.
B. The changes in information risk will impact critical business processes. The risk practitioner should report this to management so that management is able make informed risk response decisions.
C. Gaining support for new countermeasures is not a primary reason to report changes in information risk to senior management. Not all significant changes require new countermeasures.
D. Recalculation of asset values is not a primary reason to report changes in information risk to senior management. Senior management generally understands the importance of critical assets and does not wait for significant risk to reconsider the asset value.

R4-51 What is the **MOST** essential attribute of an effective key risk indicator (KRI)?

A. The KRI is accurate and reliable.
B. The KRI is predictive of a risk event.
C. The KRI provides quantitative metrics.
D. The KRI indicates required action.

B is the correct answer.

Justification:
A. Key risk indicators (KRIs) are usually indicators that risk is developing and typically are neither accurate nor reliable in the sense that they indicate what the actual risk is.
B. A KRI should indicate that a risk is developing or changing to show that investigation is needed to determine the nature and extent of a risk.
C. KRIs typically do not provide quantitative metrics about risk.
D. KRIs will not indicate that any particular action is required other than to investigate further.

R4-52 A company has set the unacceptable error level at 10 percent. Which of the following tools can be used to trigger a warning when the error level reaches eight percent?

A. A fault tree analysis
B. Statistical process control (SPC)
C. A key performance indicator (KPI)
D. A failure modes and effects analysis (FMEA)

C is the correct answer.

Justification:
A. A fault tree analysis is used to identify the sources of a risk, but not the measurement of risk.
B. Statistical process control (SPC) is used for statistical process control, not performance management.
C. A key performance indicator (KPI) is a tool that will show a performance change indication. A KPI is a measure that determines how well the process is performing in enabling the goal to be reached.
D. A failure modes and effects analysis (FMEA) is a tool that is used for failure analysis, not performance management.

R4-53 When would a risk professional ideally perform a complete enterprisewide threat analysis?

A. On a yearly basis
B. When malware is detected
C. When regulatory requirements change
D. Following a security incident

A is the correct answer.

Justification:

A. **A complete threat analysis would be performed on a yearly basis, and it can be broken down into monthly or quarterly increments, if desired.**
B. Malware could be detected all the time in a filtering system and not do any harm. Even if malware did infect the enterprise, this is not the time to perform a complete threat analysis.
C. Changes in regulations would mandate an analysis of the effect on information and information systems, but not a complete threat analysis.
D. Following a security incident, the risk professional will want to perform a threat analysis of the affected areas, but perhaps not a complete threat analysis.

R4-54 Risk monitoring provides timely information on the actual status of the enterprise with regard to risk. Which of the following choices provides an overall risk status of the enterprise?

A. Risk management
B. Risk analysis
C. Risk appetite
D. Risk profile

D is the correct answer.

Justification:

A. Risk management is the coordinated activities to direct and control an enterprise with regard to risk.
B. Risk analysis is the analysis of risk at a point in time and is not updated via multiple sources with the current actual risk status of the entire enterprise. The initial steps of risk management are analyzing the value of assets to the business, identifying threats to those assets and evaluating how vulnerable each asset is to those threats. Risk management often involves an evaluation of the probable frequency of a particular event as well as the probable impact of that event.
C. Risk appetite is the amount of risk, on a broad level, that an entity is willing to accept in pursuit of its mission and is not a current status of overall risk.
D. **The risk profile provides the current overall portfolio of the identified risk to which the enterprise is exposed. Because the profile is kept updated with evolving and new risk, it provides the enterprise's current risk status.**

R4-55 Reliability of a key risk indicator (KRI) would indicate that the metric:

A. performs within the appropriate thresholds.
B. tests the target at predetermined intervals.
C. flags exceptions every time they occur.
D. initiates corrective action.

C is the correct answer.

Justification:

A. Sensitivity of the key risk indicator (KRI) relates to the variation from a defined state that the indicator will allow before it flags an exception. The smaller the variation, the more sensitive the KRI. While sensitivity may affect the reliability of the KRI, sensitivity itself is not sufficient to determine reliability.
B. Testing the target at predetermined intervals relates to the frequency of the KRI. While frequency may affect the reliability of the KRI, frequency itself is not sufficient to determine reliability.
C. KRIs that are reporting on the data points that cannot be controlled by the enterprise, or are not alerting management at the correct time to an adverse condition, must be adjusted (optimized) to be more precise, more relevant or more accurate. Flagging exceptions every time they occur indicates the reliability of the KRI.
D. Reliability does not initiate corrective action; it means that there is a high correlation with the risk and is a good predictor or outcome measure.

R4-56 How can an enterprise determine the aggregated risk from several sources?

A. Through a security information and event management (SIEM) system
B. Through a fault tree analysis
C. Through a failure modes and effects analysis
D. Through a business impact analysis (BIA)

A is the correct answer.

Justification:

A. A security information and event management (SIEM) system will gather incident activity from several locations and prepare reports from risk trends and correlated events.
B. A fault tree analysis will examine all of the factors that could lead to a risk, but will not correlate or aggregate risk from several sources.
C. A failure modes and effects analysis will examine the sequence of events and impacts of an incident, but will not aggregate risk data.
D. A business impact analysis (BIA) will provide an understanding for a particular business unit; however, it will not be a means of determining aggregated risk from several sources.

R4-57 What is the **MOST** important criterion when reviewing information security controls?

A. To provide assurance to management of control monitoring
B. To ensure that the controls are effectively addressing risk
C. To review the impact of the controls on business operations and performance
D. To establish a baseline as a benchmark for future tests

B is the correct answer.

Justification:
A. It is important to inform management of the monitoring and testing of controls, but that is not the primary purpose of a control.
B. The primary purpose of a control is to ensure that it is effectively addressing the risk for which the control was selected and implemented.
C. The impact of the control on performance is secondary to the requirement to ensure that the control is properly addressing risk.
D. Providing a benchmark for future tests is not the primary purpose of a control review.

R4-58 Control objectives are useful to risk professionals because they provide the basis for understanding the:

A. techniques for securing information for a given risk.
B. information security policies, procedures and standards.
C. control best practices relevant to a specific entity.
D. desired outcome of implementing specific control procedures.

D is the correct answer.

Justification:
A. IT control objectives will not provide the technique for securing information for a given risk. The technique for security information for a given risk will be determined by selecting controls and defining the technique in the way in which it will work in the control environment.
B. To understand security policies, procedures and standards better, one has to better understand the business, the risk involved in various processes and how these policies will manage risk. However, IT control objectives in themselves will not help one to understand the security policies, procedures and standards.
C. The IT control objectives do not mandate best practice; they help establish the need for and the desired outcome of a control.
D. IT control objectives define the main purpose or objective for an IT control and help implement specific control procedures.

R4-59 Which of the following activities should a risk professional perform to determine whether firewall deployments are deviating from the enterprise's information security policy?

A. Review the firewall parameter settings.
B. Review the firewall intrusion prevention system (IPS) logs.
C. Review the firewall hardening procedures.
D. Analyze the firewall log file for recent attacks.

A is the correct answer.

Justification:

A. Firewall parameter settings will tie with the configurations which are linked to the governing security policy. So if the parameter settings are different than what policy states/requires, then there is a deviation.

B. Reviewing the intrusion protection system (IPS) logs may point out, to some extent, what packets were not blocked at the firewall level. To determine whether the firewall is compliant with the enterprise's security policy, one has to review the parameters such as firewall rules for traffic management, connectivity and firewall configurations.

C. Reviewing firewall hardening procedures will help one understand what was expected for security of the firewall, but without review of the actual firewall settings, one cannot establish whether the deployments deviate from the enterprise's security policy.

D. There can be attacks on the firewall for which the enterprise may not have formally defined rules in the security policy; analyzing firewall logs for recent attacks does not imply that there is a firewall policy deviation.

R4-60 Which of the following choices is the **MOST** important critical success factor (CSF) of implementing a risk-based approach to the system development life cycle (SDLC)?

A. Existence of a risk management framework
B. Defined risk mitigation strategies
C. Compliance with the change management process
D. Adequate involvement of business representatives

D is the correct answer.

Justification:

A. The existence of a risk management framework does not necessarily ensure compliance and success during the system development life cycle (SDLC).

B. Understanding the defined risk mitigation strategies will help the enterprise manage risk effectively; however, adequate involvement of business representatives is still required.

C. Although compliance with the change management process is a critical success factor (CSF) for system development, it is not the most important of the options provided.

D. A CSF for system development is the adequate involvement of business representatives, including management, users, quality assurance, IT, privacy, legal, audit, regulatory affairs or compliance teams in high-risk regulatory situations.

R4-61 Monitoring has flagged a security exception. What is the **MOST** appropriate action?

A. Escalate the exception.
B. Update the risk register.
C. Activate the risk response plan.
D. Validate the exception.

D is the correct answer.

Justification:

A. The escalation to management should not occur until more is known about the situation, and even then only if it is outside the security manager's scope to address the issue.
B. The risk register should be updated after the exception has been validated.
C. The risk response plan will not be activated until the exception has been validated and the response has been approved by management.
D. Before any other action is taken, the security manager should ensure that the exception identified by monitoring is not a false positive.

R4-62 Which of the following criteria is **MOST** essential for the effectiveness of operational metrics?

A. Relevance to the recipient
B. Timeliness of the reporting
C. Accuracy of the measurement
D. Cost of obtaining the metrics

A is the correct answer.

Justification:

A. Unless the metric is relevant to the recipient and the recipient understands what the metric means and what action to take, if any, all other criteria are of little importance.
B. Timeliness of reporting is important, but secondary to relevance.
C. A high degree of accuracy is not essential as long as the metric is reliable and indications are within an acceptable range.
D. Cost is always a consideration, but secondary to relevance.

R4-63 Which of the following is the **MOST** appropriate metric to measure how well the information security function is managing the administration of user access?

A. Elapsed time to suspend accounts of terminated users
B. Elapsed time to suspend accounts of users transferring
C. Ratio of actual accounts to actual end users
D. Percent of accounts with configurations in compliance

D is the correct answer.

Justification:

A. Elapsed time to suspend accounts of terminated users is only part of the picture and does not address the volume of requests.
B. Elapsed time to suspend accounts of users transferring is only part of the picture and does not address the volume of requests.
C. The ratio of actual accounts to actual end users does not indicate much in terms of how well security is administered.
D. The percent of accounts with configurations in compliance is the best measure of how well the administration is being managed because this shows the overall impact.

R4-64 Which of the following **BEST** assists in the proper design of an effective key risk indicator (KRI)?

A. Generating the frequency of reporting cycles to report on the risk
B. Preparing a business case that includes the measurement criteria for the risk
C. Conducting a risk assessment to provide an overview of the key risk
D. Documenting the operational flow of the business from beginning to end

D is the correct answer.

Justification:

A. Generating the frequency of reporting for the key risk indicator (KRI) means nothing if the KRI is not designed.
B. A proper business case describes what is going to be done, why it is worth doing, how it will be accomplished and what resources will be required. It will not document the data points, structures or any other needed data for designing a KRI.
C. A risk assessment is the determination of a value of risk related to some situation and a recognized threat. While it contributes somewhat to the design of the KRI, there still is a need for additional information.
D. Prior to starting to design the KRI, a risk manager must understand the end-to-end operational flow of the respective business. This gives insight into the detailed processes, data flows, decision-making processes, acceptable levels of risk for the business, etc., which in turn give the risk manager the ability to apply top and bottom levels for the KRI.

R4-65 One way to determine control effectiveness is by determining:

A. the test results of intended objectives.
B. whether it is preventive, detective or compensatory.
C. the capability of providing notification of failure.
D. the evaluation and analysis of reliability.

A is the correct answer.

Justification:

A. Control effectiveness requires a process to verify that the control process worked as intended. Examples such as dual-control or dual-entry bookkeeping provide verification and assurance that the process operated as intended.
B. The type of control is not relevant.
C. Notification of failure is not determinative of control strength.
D. Reliability is not an indication of control strength; weak controls can be highly reliable, even if they are ineffective controls.

R4-66 Implementing continuous monitoring controls is the **BEST** option when:

A. legislation requires strong information security controls.
B. incidents may have a high impact and frequency.
C. incidents may have a high impact, but low frequency.
D. e-commerce is a primary business driver.

B is the correct answer.

Justification:
A. Regulations and legislation that require tight IT security measures focus on requiring enterprises to establish an IT security governance structure that manages IT security with a risk-based approach, so each organization decides which kinds of controls are implemented. Continuous monitoring is not necessarily a requirement.
B. Because they are expensive, continuous monitoring control initiatives are used in areas where the risk is at its greatest level. These areas have a high impact and frequency of occurrence.
C. Measures such as contingency planning are commonly used when incidents rarely happen, but have a high impact each time they happen. Continuous monitoring is unlikely to be necessary.
D. Continuous control monitoring initiatives are not needed in all e-commerce environments. There are some e-commerce environments where the impact of incidents is not high enough to support the implementation of this kind of initiative.

R4-67 When the key risk indicator (KRI) for the IT change management process reaches its threshold, a risk practitioner should **FIRST** report this to the:

A. business owner.
B. chief information security officer (CISO).
C. help desk.
D. incident response team.

A is the correct answer.

Justification:
A. Reporting to the business owners first is the most appropriate action because they own the risk and determine the risk response.
B. Reporting to the chief information security officer (CISO) is important, but is not as critical as reporting to the business owners.
C. Reporting to the help desk is not appropriate when reporting on risk. The report must go to the business owners because they own the risk and determine the risk response.
D. Reporting to the incident response team is not appropriate when reporting on risk. The report must go to the business owners because they own the risk and determine the risk response.

R4-68 Which of the following **MUST** be included when developing metrics to identify and monitor the control life cycle?

A. Thresholds that identify when controls no longer provide the intended value
B. Customized reports of the metrics for key stakeholders
C. A description of the methods and practices used to develop the metrics
D. Identification of a repository where metrics will be maintained and stored

A is the correct answer.

Justification:

A. Metrics used to monitor the control life cycle require thresholds to identify when controls are no longer providing their intended value, which ensures that the enterprise is aware and can take appropriate action. Without this information, an enterprise may be under the impression that ineffective controls are still effective and do not need to be adjusted or retired.
B. All of the other choices listed are valuable, but not required as part of the metric and measurement development process for life cycle controls.
C. While a description of how a specific metric has been developed may be useful for knowledge transfer and to improve the metric over time, it is not a requirement.
D. A repository where metrics are maintained or stored may be useful in larger enterprises; however, for smaller enterprises there may be no need to create a formal repository.

R4-69 The **MOST** important objective of regularly testing information system controls is to:

A. identify design flaws, failures and redundancies.
B. provide the necessary evidence to support management assertions.
C. assess the control risk and formulate an opinion on the level of reliability.
D. evaluate the need for a risk assessment and indicate the corrective action(s) to be taken, where applicable.

A is the correct answer.

Justification:

A. This choice is the best statement because it contains the necessary activities needed to ensure that the control is designed correctly and is operating effectively and efficiently during the production phase.
B. This activity is performed after the completion of an assessment or audit of the information system control.
C. This activity is primarily performed during the design phase of the information system control.
D. Risk assessments do not depend on testing of controls.

R4-70 What is the **MOST** important factor in the success of an ongoing information security monitoring program?

A. Logs that capture all network and application traffic for later analysis
B. Staff who are qualified and trained to execute their responsibilities
C. System components all have up-to-date patches
D. A security incident and event management (SIEM) system is in place

B is the correct answer.

Justification:

A. While capturing traffic logs is important, capturing the logs in itself is not a value-added activity unless the logs are being reviewed by a qualified person. Without logs it can be difficult to really "know" what is happening; however, the logs are of little value unless they are being reviewed by a qualified person.

B. Information security monitoring requires the gathering and analysis of data and reporting the results to management. This requires staff who are trained in using the tools, generating the data requests, performing the analysis and being able to communicate effectively. Not having staff with adequate training will result in a monitoring effort that may be inaccurate, incomplete, or may miss critical trends.

C. Keeping patch levels up to date is a core operational task that helps ensure that the systems remain protected from newly discovered vulnerabilities. However, this is not a critical part of a successful security monitoring program.

D. Tools are only as good as the staff who manages the tools. There are some excellent security incident and event management (SIEM) tools that will gather data from many points, correlate the data and generate excellent reports on the activity on the system—if the tool is set up and managed correctly.

R4-71 What role does the risk professional have in regard to the IS control monitoring process? The risk professional:

A. maintains and operates IS controls.
B. approves the policies for IS control monitoring.
C. determines the frequency of control testing by internal audit.
D. assists in planning, reporting and scheduling tests of IS controls.

D is the correct answer.

Justification:

A. The risk professional is a consultative position and should ensure that tests are being conducted and reported on and that mitigation efforts are conducted as necessary. The risk professional is not accountable for the maintenance and operation of the controls.

B. Policy approval is a governance function and not part of the role of the risk professional.

C. The risk professional does play a consultative role for audit and other executive functions, but does not determine the frequency of control testing.

D. The risk professional plays a key role in scheduling, supervising and reporting on risk. This includes the responsibility for working with the testing teams.

R4-72 What is the **MOST** important reason for periodically testing controls?

A. To meet regulatory requirements
B. To meet due care requirements
C. To ensure that control objectives are met
D. To achieve compliance with standard policy

C is the correct answer.

Justification:

A. The testing of controls is important for more than just compliance with regulatory requirements. Many controls are not related to a regulatory requirement.
B. Periodically testing controls does not help meet due care requirements. Due care is what a reasonable person of similar competency would do under similar circumstances. The testing of the controls is demonstration of due diligence, not due care. Due care puts the control in place, due diligence tests that the control is working.
C. Periodically testing controls ensures that controls continue to meet control objectives.
D. Compliance with policy is not the most important factor for periodically testing controls.

R4-73 Which of the following measures is **MOST** effective against insider threats to confidential information?

A. Audit trail monitoring
B. A privacy policy
C. Role-based access control (RBAC)
D. Defense in depth

C is the correct answer.

Justification:

A. Audit trail monitoring is a detective control, which is "after the fact."
B. A privacy policy is not relevant to this risk.
C. Role-based access control (RBAC) provides access according to business needs; therefore, it reduces unnecessary access rights and enforces accountability.
D. The primary focus of defense in depth is external threats.

R4-74 A well-known hacking group has publicly stated they will target a company. What is the risk professional's **FIRST** action?

A. Advise IT management about the threat.
B. Inform all employees about the threat.
C. Contact law enforcement officials about the threat.
D. Inform senior management about the threat.

D is the correct answer.

Justification:

A. Only critical members of the IT management team should be notified of the threat.
B. Information should be given on a need-to-know basis; all employees do not need to know of a potential threat to the company.
C. Contacting law enforcement at this time is premature, although law enforcement may need to be contacted in the future with management approval.
D. All senior management needs to be aware of the threat so that they can be prepared if an incident takes place.

R4-75 What is the **BEST** approach to determine whether existing security control management meets the organizational needs?

A. Perform a process maturity assessment.
B. Perform a control self-assessment (CSA).
C. Review security logs for trends or issues.
D. Compare current and historical security test results.

A is the correct answer.

Justification:

A. **A process maturity assessment can be used to determine the presence of the control as well as the reliable operation and maintenance of the control, and determine any gaps between the desired and current state of the control.**
B. Control self-assessments (CSAs) are a valuable tool to monitor controls on an ongoing basis, but will not indicate the maturity of the security control management process.
C. Logs will record what has happened, but they will not indicate whether the configurations used to create the logs are incorrect.
D. Running test data through the system and comparing to previous results will show whether the effectiveness of the controls has changed, but will not indicate whether the controls are effectively being maintained or are effective to mitigate new risk.

R4-76 Which of the following practices **BEST** mitigates the risk associated with outsourcing a business function?

A. Performing audits to verify compliance with contract requirements
B. Requiring all vendor staff to attend annual awareness training sessions
C. Retaining copies of all sensitive data on internal systems
D. Reviewing the financial records of the vendor to verify financial soundness

A is the correct answer.

Justification:

A. **Regular audits verify that the vendor is compliant with contract requirements.**
B. Requiring the vendor staff to attend annual awareness sessions is not usually part of an outsourcing contract, although it may be a good idea.
C. Keeping copies of all sensitive data is an unnecessary expenditure and may result in errors or inconsistencies with data stored at the vendor site. In addition, duplicating sensitive data means that the company is now liable for protecting data in two or more locations and increases the possibility of inappropriate access and/or data leakage.
D. Although it is common practice to review financial solvency before selecting a vendor to ensure that the vendor functions without the threat of liquidation for the foreseeable future, reviewing solvency is not the best practice to address the risk related to outsourcing an IT or business function.

Page intentionally left blank

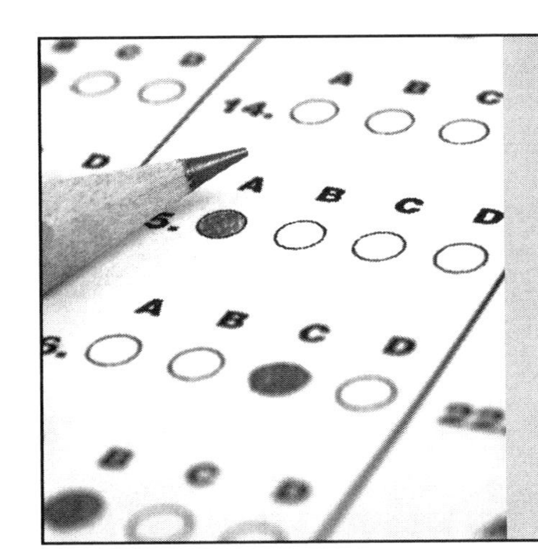

POSTTEST

If you wish to take a posttest to determine strengths and weaknesses, the Sample Exam begins on page 181 and the posttest answer sheet begins on page 207. You can score your posttest with the Sample Exam Answer and Reference Key on page 203.

Page intentionally left blank

SAMPLE EXAM

1. Management wants to ensure that IT is successful in delivering against business requirements. Which of the following **BEST** supports that effort?

   A. An internal control system or framework
   B. A cost-benefit analysis
   C. A return on investment (ROI) analysis
   D. A benchmark process

2. Which of the following is the **MOST** important reason for conducting security awareness programs throughout an enterprise?

   A. Reducing the risk of a social engineering attack
   B. Training personnel in security incident response
   C. Informing business units about the security strategy
   D. Maintaining evidence of training records to ensure compliance

3. Who is accountable for business risk related to IT?

   A. The chief information officer (CIO)
   B. The chief financial officer (CFO)
   C. Users of IT services—the business
   D. The chief architect

4. Which of the following should be of **MOST** concern to a risk practitioner?

   A. Failure to notify the public of an intrusion
   B. Failure to notify the police of an attempted intrusion
   C. Failure to internally report a successful attack
   D. Failure to examine access rights periodically

5. An excessive number of standard workstation images can be categorized as a key risk indicator (KRI) for:

   A. change management.
   B. configuration management.
   C. IT operations management.
   D. data management.

6. Which of the following is **MOST** beneficial to the improvement of an enterprise's risk management process?

   A. Key risk indicators (KRIs)
   B. External benchmarking
   C. The latest risk assessment
   D. A maturity model

7. Once a risk assessment has been completed, the documented test results should be:

   A. destroyed.
   B. retained.
   C. summarized.
   D. published.

8. A risk assessment process that uses likelihood and impact in calculating the level of risk is a:

A. qualitative process.
B. failure modes and effects analysis (FMEA).
C. fault tree analysis.
D. quantitative process.

9. A substantive test to verify that tape library inventory records are accurate is:

A. determining whether bar code readers are installed.
B. conducting a physical count of the tape inventory.
C. checking whether receipts and issues of tapes are accurately recorded.
D. determining whether the movement of tapes is authorized.

10. Which of the following tools aids management in determining whether a project should continue based on scope, schedule and cost? Analysis of:

A. earned value management.
B. the function point.
C. the Gantt chart.
D. the program evaluation and review technique (PERT).

11. Risk scenarios enable the risk assessment process because they:

A. cover a wide range of potential risk.
B. minimize the need for quantitative risk analysis techniques.
C. segregate IT risk from business risk for easier risk analysis.
D. help estimate the frequency and impact of risk.

12. Which of the following **MOST** likely indicates that a customer data warehouse should remain in-house rather than be outsourced to an offshore operation?

A. The telecommunications costs may be much higher in the first year.
B. Privacy laws may prevent a cross-border flow of information.
C. Time zone differences may impede communications between IT teams.
D. Software development may require more detailed specifications.

13. What is the **MOST** important factor in the success of an ongoing information security monitoring program?

A. Logs that capture all network and application traffic for later analysis
B. Staff who are qualified and trained to execute their responsibilities
C. System components all have up-to-date patches
D. A security incident and event management (SIEM) system is in place

14. Which of the following measures is **MOST** effective against insider threats to confidential information?

A. Audit trail monitoring
B. A privacy policy
C. Role-based access control (RBAC)
D. Defense in depth

15. The **BEST** method for detecting and monitoring a hacker's activities without exposing information assets to unnecessary risk is to utilize:

A. firewalls.
B. bastion hosts.
C. honeypots.
D. screened subnets.

16. Reliability of a key risk indicator (KRI) would indicate that the metric:

A. performs within the appropriate thresholds.
B. tests the target at predetermined intervals.
C. flags exceptions every time they occur.
D. initiates corrective action.

17. Which of the following **BEST** describes the information needed for each risk on a risk register?

A. Various risk scenarios with their date, description, impact, probability, risk score, mitigation action and owner
B. Various risk scenarios with their date, description, risk score, cost to remediate, communication plan and owner
C. Various risk scenarios with their date, description, impact, cost to remediate and owner
D. Various activities leading to risk management planning

18. When requesting information for an e-discovery, an enterprise learned that their email cloud provider was never contracted to back up the messages even though the company's email retention policy explicitly states that all emails are to be saved for three years. Which of the following would have **BEST** safeguarded the company from this outcome?

A. Providing the contractor with the record retention policy up front
B. Validating the company policies to the provider's contract
C. Providing the contractor with the email retention policy up front
D. Backing up the data on the company's internal network nightly

19. Which of the following is the **BEST** method to ensure the overall effectiveness of a risk management program?

A. Assignment of risk within the enterprise
B. Comparison of the program results with industry standards
C. Participation by applicable members of the enterprise
D. User assessment of changes in risk

20. When a significant vulnerability is discovered in the security of a critical web server, immediate notification should be made to the:

A. development team to remediate.
B. data owners to mitigate damage.
C. system owner to take corrective action.
D. incident response team to investigate.

21. The preparation of a risk register begins in which risk management process?

A. Risk response planning
B. Risk monitoring and control
C. Risk management planning
D. Risk identification

22. The **BEST** reason to implement a maturity model for risk management is to:

A. permit alignment with business objectives.
B. help improve governance and compliance.
C. ensure that security controls are effective.
D. enable continuous improvement.

23. The goal of IT risk analysis is to:

A. enable the alignment of IT risk management with enterprise risk management (ERM).
B. enable the prioritization of risk responses.
C. satisfy legal and regulatory compliance requirements.
D. identify known threats and vulnerabilities to information assets.

24. Which of the following activities provides the **BEST** basis for establishing risk ownership?

A. Documenting interdependencies between departments
B. Mapping identified risk to a specific business process
C. Referring to available RACI charts
D. Distributing risk equally among all asset owners

25. Business stakeholders and decision makers reviewing the effectiveness of IT risk responses would **PRIMARILY** validate whether:

A. IT controls eliminate the risk in question.
B. IT controls are continuously monitored.
C. IT controls achieve the desired objectives.
D. IT risk indicators are formally documented.

26. As part of risk monitoring, the administrator of a two-factor authentication system identifies a trusted independent source indicating that the algorithm used for generating keys has been compromised. The vendor of the authentication system has not provided further information. Which of the following is the **BEST** initial course of action?

A. Wait for the vendor to formally confirm the breach and provide a solution.
B. Determine and implement suitable compensating controls.
C. Identify all systems requiring two-factor authentication and notify their business owners.
D. Disable the system and rely on the single-factor authentication until further information is received.

27. An enterprise expanded operations into Europe, Asia and Latin America. The enterprise has a single-version, multiple-language employee handbook last updated three years ago. Which of the following is of **MOST** concern?

A. The handbook may not have been correctly translated into all languages.
B. Newer policies may not be included in the handbook.
C. Expired policies may be included in the handbook.
D. The handbook may violate local laws and regulations.

28. After a risk assessment study, a bank with global operations decided to continue doing business in certain regions of the world where identity theft is widespread. To **MOST** effectively deal with the risk, the business should:

A. implement monitoring techniques to detect and react to potential fraud.
B. make the customer liable for losses if the customer fails to follow the bank's advice.
C. increase its customer awareness efforts in those regions.
D. outsource credit card processing to a third party.

29. Which of the following provides the formal authorization on user access?

A. Database administrator
B. Data owner
C. Process owner
D. Data custodian

30. A risk practitioner has collected several IT-related key risk indicators (KRIs) related for the core financial application. These would **MOST** likely be reported to:

A. stakeholders.
B. the IT administrator group.
C. the finance department.
D. senior management.

31. After the completion of a risk assessment, it is determined that the cost to mitigate the risk is much greater than the benefit to be derived. A risk practitioner should recommend to business management that the risk be:

A. treated.
B. terminated.
C. accepted.
D. transferred.

32. Risk monitoring provides timely information on the actual status of the enterprise with regard to risk. Which of the following choices provides an overall risk status of the enterprise?

A. Risk management
B. Risk analysis
C. Risk appetite
D. Risk profile

33. It is **MOST** important that risk appetite be aligned with business objectives to ensure that:

A. resources are directed toward areas of low risk tolerance.
B. major risk is identified and eliminated.
C. IT and business goals are aligned.
D. the risk strategy is adequately communicated.

34. Risk assessment techniques should be used by a risk practitioner to:

A. maximize the return on investment (ROI).
B. provide documentation for auditors and regulators.
C. justify the selection of risk mitigation strategies.
D. quantify the risk that would otherwise be subjective.

35. Which of the following is the **GREATEST** benefit of a risk-aware culture?

A. Issues are escalated when suspicious activity is noticed.
B. Controls are double-checked to anticipate any issues.
C. Individuals communicate with peers for knowledge sharing.
D. Employees are self-motivated to learn about costs and benefits.

36. Which of the following **BEST** identifies changes in an enterprise's risk profile?

A. The risk register
B. Risk classification
C. Changes in risk indicator thresholds
D. Updates to the control inventory

37. There is an increase in help desk call levels because the vendor hosting the human resources (HR) self-service portal has reduced the password expiration from 90 to 30 days. The corporate password policy requires password expiration after 60 days and HR is unaware of the change. The risk practitioner should **FIRST**:

A. formally investigate the cause of the unauthorized change.
B. request the service provider reverse the password expiration period to 90 days.
C. initiate a request to strengthen the corporate password expiration requirement to 30 days.
D. notify employees of the change in password expiration period.

38. Despite a comprehensive security awareness program annually undertaken and assessed for all staff and contractors, an enterprise has experienced a breach through a spear phishing attack. What is the **MOST** effective way to improve security awareness?

A. Review the security awareness program and improve coverage of social engineering threats.
B. Launch a disciplinary process against the people who leaked the information.
C. Perform a periodic social engineering test against all staff and communicate summary results to the staff.
D. Implement a data loss prevention system that automatically points users to corporate policies.

39. Which of the following is the **MOST** important information to include in a risk management strategic plan?

A. Risk management staffing requirements
B. The risk management mission statement
C. Risk mitigation investment plans
D. The current state and desired future state

40. Which of the following is **MOST** important to determine when defining risk management strategies?

A. Risk assessment criteria
B. IT architecture complexity
C. An enterprise disaster recovery plan (DRP)
D. Organizational objectives

41. Which of the following **BEST** ensures that identified risk is kept at an acceptable level?

A. Reviewing of the controls periodically, according to the risk action plan
B. Listing each risk as a separate entry in the risk register
C. Creating a separate risk register for every department
D. Maintaining a key risk indicator (KRI) for assets in the risk register

42. The board of directors of a one-year-old start-up company has asked their chief information officer (CIO) to create all of the enterprise's IT policies and procedures, which will be managed and approved by the IT steering committee. The IT steering committee will make all of the IT decisions for the enterprise, including those related to the technology budget. The IT steering committee will be **BEST** represented by:

A. members of the executive board.
B. high-level members of the IT department.
C. IT experts from outside of the enterprise.
D. key members from each department.

43. Risk scenarios should be created **PRIMARILY** based on which of the following?

A. Input from senior management
B. Previous security incidents
C. Threats that the enterprise faces
D. Results of the risk analysis

44. Which of the following is **MOST** critical when system configuration files for a critical enterprise application system are being reviewed?

A. Configuration files are frequently changed.
B. Changes to configuration files are recorded.
C. Access to configuration files is not restricted.
D. Configuration values do not impact system efficiency.

45. Due to changes in the IT environment, the disaster recovery plan of a large enterprise has been modified. What is the **GREATEST** benefit of testing the new plan?

A. To ensure that the plan is complete
B. To ensure that the team is trained
C. To ensure that all assets have been identified
D. To ensure that the risk assessment was validated

46. It is **MOST** important for risk mitigation to:

A. eliminate threats and vulnerabilities.
B. reduce the likelihood of risk occurrence.
C. reduce risk within acceptable cost.
D. reduce inherent risk to zero.

47. Which of the following is a **PRIMARY** role of the system owner during the accreditation process? The system owner:

A. reviews and approves the security plan supporting the system.
B. selects and documents the security controls for the system.
C. assesses the security controls in accordance with the assessment procedures.
D. determines whether the risk to the business is acceptable.

48. What is the **MOST** effective method to evaluate the potential impact of legal, regulatory and contractual requirements on business objectives?

A. A compliance-oriented gap analysis
B. Interviews with business process stakeholders
C. A mapping of compliance requirements to policies and procedures
D. A compliance-oriented business impact analysis (BIA)

49. Which type of cost incurred is used when leveraging existing network cabling for an IT project?

A. Indirect cost
B. Infrastructure cost
C. Project cost
D. Maintenance cost

50. Which of the following **MOST** enables risk-aware business decisions?

A. Robust information security policies
B. An exchange of accurate and timely information
C. Skilled risk management personnel
D. Effective process controls

51. A review of an enterprise's IT projects finds that projects frequently go over time or budget by nearly 10 percent. On review, management advises the risk practitioner that a deviation of 15 percent is acceptable. This is an example of:

A. risk avoidance.
B. risk tolerance.
C. risk acceptance.
D. risk mitigation.

52. Which type of risk assessment methods involves conducting interviews and using anonymous questionnaires by subject matter experts?

A. Quantitative
B. Probabilistic
C. Monte Carlo
D. Qualitative

53. The board of directors wants to know the financial impact of specific, individual risk scenarios. What type of approach is **BEST** suited to fulfill this requirement?

A. Delphi method
B. Quantitative analysis
C. Qualitative analysis
D. Financial risk modeling

54. Which of the following is the **BEST** way to ensure that an accurate risk register is maintained over time?

A. Monitor key risk indicators (KRIs), and record the findings in the risk register.
B. Publish the risk register centrally with workflow features that periodically poll risk assessors.
C. Distribute the risk register to business process owners for review and updating.
D. Utilize audit personnel to perform regular audits and to maintain the risk register.

55. The **MOST** important external factors that should be considered in a risk assessment effort are:

A. proposed new security tools and technologies.
B. the number of viruses and other malware being developed.
C. international crime statistics and political unrest.
D. supply chain and market conditions.

56. What do different risk scenarios on the same bands/curve on a risk map indicate?

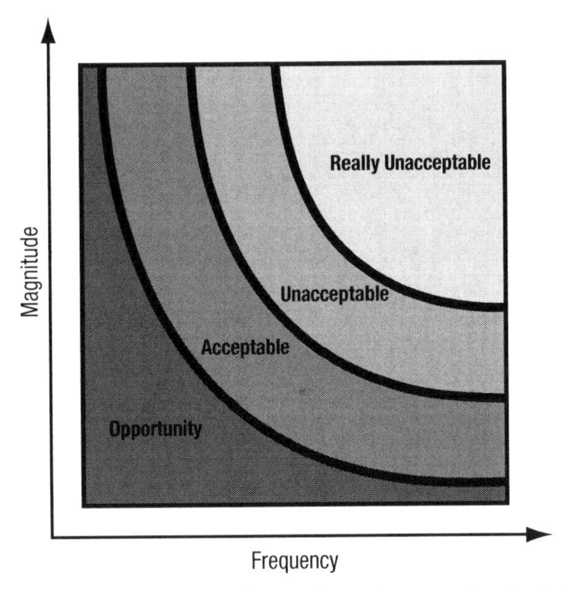

A. All risk scenarios on the same curve of a risk map have the same level of risk.
B. All risk scenarios on the same curve of a risk map have the same magnitude of impact.
C. All risk scenarios on the same curve of a risk map require the same risk response.
D. All risk scenarios on the same curve of a risk map are of the same type.

57. The **PRIMARY** purpose of adopting an enterprisewide risk management framework is to:

A. allow the flexibility to adjust the risk response strategy throughout the enterprise.
B. centralize the responsibility for the maintenance of the risk response program.
C. enable a consistent approach to risk response throughout the enterprise.
D. avoid higher costs for risk reduction and audit strategies throughout the enterprise.

58. During an internal risk assessment in a global enterprise, a risk manager notes that local management has proactively mitigated some of the high-level risk related to the global purchasing process. This means that:

A. the local management is now responsible for the risk.
B. the risk owner is the corporate chief risk officer (CRO).
C. the risk owner is the local purchasing manager.
D. corporate management remains responsible for the risk.

59. An enterprise is hiring a consultant to help determine the maturity level of the risk management program. The **MOST** important element of the request for proposal (RFP) is the:

A. sample deliverable.
B. past experience of the engagement team.
C. methodology used in the assessment.
D. references from other organizations.

60. A company is confident about the state of its organizational security and compliance program. Many improvements have been made since the last security review was conducted one year ago. What should the company do to evaluate its current risk profile?

A. Review previous findings and ensure that all issues have been resolved.
B. Conduct follow-up audits in areas that were found deficient in the previous review.
C. Monitor the results of the key risk indicators (KRIs) and use those to develop targeted assessments.
D. Perform a new enterprise risk assessment using an independent expert.

61. Which of the following assessments of an enterprise's risk monitoring process will provide the **BEST** information about its alignment with industry-leading practices?

A. A capability assessment by an outside firm
B. A self-assessment of capabilities
C. An independent benchmark of capabilities
D. An internal audit review of capabilities

62. The cost of mitigating a risk should not exceed the:

A. expected benefit to be derived.
B. annual loss expectancy (ALE).
C. value of the physical asset.
D. cost to exploit the weakness.

63. Which of the following is **MOST** important when considering the risk appetite of an enterprise?

A. The capacity of the enterprise to absorb loss
B. The definition of responsibilities for risk management
C. The line of business and the typical risk of the industry
D. The culture and predisposition toward risk taking

64. How does an enterprise **BEST** ensure that developers do not have access to implement changes to production applications?

A. The enterprise must ensure that development staff does not have access to executable code.
B. The enterprise must have segregation of duties between application development and operations.
C. The enterprise system development life cycle (SDLC) must be enforced to require segregation of duties.
D. The enterprise's change management process must be enforced for all but emergency changes.

65. A database administrator notices that the externally hosted, web-based corporate address book application requires users to authenticate, but that the traffic between the application and users is not encrypted. The **MOST** appropriate course of action is to:

A. notify the business owner and the security manager of the discovery and propose an addition to the risk register.
B. contact the application administrators and request that they enable encryption of the application's web traffic.
C. alert all staff about the vulnerability and advise them not to log on from public networks.
D. accept that current controls are suitable for nonsensitive business data.

66. The **MOST** important objective of regularly testing information system controls is to:

A. identify design flaws, failures and redundancies.
B. provide the necessary evidence to support management assertions.
C. assess the control risk and formulate an opinion on the level of reliability.
D. evaluate the need for a risk assessment and indicate the corrective action(s) to be taken, where applicable.

67. Prior to releasing an operating system security patch into production, a leading practice is to have the patch:

A. applied simultaneously to all systems.
B. procured from an approved vendor.
C. tested in a preproduction test environment.
D. approved by business stakeholders.

68. Previously accepted risk should be:

A. reassessed periodically because the risk can be escalated to an unacceptable level due to revised conditions.
B. removed from the risk log once it is accepted.
C. accepted permanently because management has already spent resources (time and labor) to conclude that the risk level is acceptable.
D. avoided next time because risk avoidance provides the best protection to the enterprise.

69. An enterprise has outsourced the majority of its IT department to a third party whose servers are in a foreign country. Which of the following is the **MOST** critical security consideration?

A. A security breach notification may get delayed due to the time difference.
B. Additional network intrusion detection sensors should be installed, resulting in additional cost.
C. The enterprise could be unable to monitor compliance with its internal security and privacy guidelines.
D. Laws and regulations of the country of origin may not be enforceable in the foreign country.

70. Which of the following is the **BEST** approach when conducting an IT risk awareness campaign?

A. Provide technical details on exploits.
B. Provide common messages tailored for different groups.
C. Target system administrators and help desk staff.
D. Target senior managers and business process owners.

71. Who should be accountable for the risk to an IT system that supports a critical business process?

A. IT management
B. Senior management
C. The risk management department
D. System users

72. What is the **BEST** tool for documenting the status of risk mitigation and risk ownership?

A. Risk action plans
B. Risk scenarios
C. Business impact analysis (BIA) documents
D. A risk register

73. Which of the following provides the **GREATEST** support to a risk practitioner recommending encryption of corporate laptops and removable media as a risk mitigation measure?

A. Benchmarking with peers
B. Evaluating public reports on encryption algorithm in the public domain
C. Developing a business case
D. Scanning unencrypted systems for vulnerabilities

74. Which of the following provides the **BEST** capability to identify whether controls that are in place remain effective in mitigating their intended risk?

A. A key performance indicator (KPI)
B. A risk assessment
C. A key risk indicator (KRI)
D. An audit

75. Which of the following should management use to allocate resources for risk response?

A. Audit report findings
B. Penetration test results
C. Risk analysis results
D. Vulnerability test results

76. A business case developed to support risk mitigation efforts for a complex application development project should be retained until:

A. the project is approved.
B. user acceptance of the application.
C. the application is deployed.
D. the application's end of life.

77. The sales manager of a home improvement enterprise wants to expand the services available on the enterprise's web page to include sending free promotional samples of their products to prospective clients. What is the **GREATEST** concern the risk professional would have?

A. Are there any data privacy concerns about storing client data?
B. Are there any concerns about protecting credit card or payment data?
C. Can the system be misused by a person to obtain multiple samples?
D. Will the web site be able to handle the expected volume of traffic?

78. The **GREATEST** risk to token administration is:

A. the ability to easily tamper with or steal a token.
B. the loss of network connectivity to the authentication system.
C. the inability to secure unassigned tokens.
D. the ability to generate temporary codes to log in without a token.

79. Deriving the likelihood and impact of risk scenarios through statistical methods is **BEST** described as:

A. quantitative risk analysis.
B. risk scenario analysis.
C. qualitative risk analysis.
D. probabilistic risk assessment.

80. A procurement employee notices that new printer models offered by the vendor keep a copy of all printed documents on a built-in hard disk. Considering the risk of unintentionally disclosing confidential data, the employee should:

A. proceed with the order and configure printers to automatically wipe all the data on disks after each print job.
B. notify the security manager to conduct a risk assessment for the new equipment.
C. seek another vendor that offers printers without built-in hard disk drives.
D. procure printers with built-in hard disks and notify staff to wipe hard disks when decommissioning the printer.

81. Assessing information systems risk is **BEST** achieved by:

A. using the enterprise's past actual loss experience to determine current exposure.
B. reviewing published loss statistics from comparable organizations.
C. evaluating threats associated with existing information systems assets and information systems projects.
D. reviewing information systems control weaknesses identified in audit reports.

82. Risk assessments should be repeated at regular intervals because:

A. omissions in earlier assessments can be addressed.
B. periodic assessments allow various methodologies.
C. business threats are constantly changing.
D. they help raise risk awareness among staff.

83. Which of the following environments typically represents the **GREATEST** risk to organizational security?

A. An enterprise data warehouse
B. A load-balanced, web server cluster
C. A centrally managed data switch
D. A locally managed file server

84. Risk assessments are **MOST** effective in a software development organization when they are performed:

A. before system development begins.
B. during system deployment.
C. during each stage of the system development life cycle (SDLC).
D. before developing a business case.

85. If risk has been identified, but not yet mitigated, the enterprise would:

A. record and mitigate serious risk and disregard low-level risk.
B. obtain management commitment to mitigate all identified risk within a reasonable time frame.
C. document all risk in the risk register and maintain the status of the remediation.
D. conduct an annual risk assessment, but disregard previous assessments to prevent risk bias.

86. Which of the following is minimized when acceptable risk is achieved?

A. Transferred risk
B. Control risk
C. Residual risk
D. Inherent risk

87. Which of the following objectives is the **PRIMARY** reason risk professionals conduct risk assessments?

A. To maintain the enterprise's risk register
B. To enable management to choose the right risk response
C. To provide assurance on the risk management process
D. To identify risk with the highest business impact

88. An enterprise is expanding into new nearby domestic locations (office park). Which of the following is **MOST** important for a risk practitioner to report on?

A. Competitor analysis
B. Legal and regulatory requirements
C. Political issues
D. The potential of natural disasters

89. The **MAIN** benefit of information classification is that it helps:

A. determine how information can be further labeled.
B. establish the access control matrices.
C. determine the risk tolerance level.
D. select security measures that are proportional to risk.

90. Which of the following statements **BEST** describes the value of a risk register?

A. It captures the risk inventory.
B. It drives the risk response plan.
C. It is a risk reporting tool.
D. It lists internal risk and external risk.

91. Which of the following is the **BEST** indicator of high maturity of an enterprise's IT risk management process?

A. People have appropriate awareness of risk and are comfortable talking about it.
B. Top management is prepared to invest more money in IT security.
C. Risk assessment is encouraged in all areas of IT and business management.
D. Business and IT are aligned in risk assessment and risk ranking.

92. The **PRIMARY** benefit of using a maturity model to assess the enterprise's data management process is that it:

A. can be used for benchmarking.
B. helps identify gaps.
C. provides goals and objectives.
D. enforces continuous improvement.

93. A key objective when monitoring information systems control effectiveness against the enterprise's external requirements is to:

A. design the applicable information security controls for external audits.
B. create the enterprise's information security policy provisions for third parties.
C. ensure that the enterprise's legal obligations have been satisfied.
D. identify those legal obligations that apply to the enterprise's security practices.

94. What is the **FIRST** step for a risk practitioner when an enterprise has decided to outsource all IT services and support to a third party?

A. Validate that the internal systems of the service provider are secure.
B. Enforce the regulations and standards associated with outsourcing data management for restrictions on transborder data flow.
C. Ensure that security requirements are addressed in all contracts and agreements.
D. Build a business case to perform an onsite audit of the third-party vendor.

95. Which of the following information in the risk register **BEST** helps in developing proper risk scenarios? A list of:

A. potential threats to assets.
B. residual risk on individual assets.
C. accepted risk.
D. security incidents.

96. Which of the following is the **BEST** indicator that incident response training is effective?

A. Decreased reporting of security incidents to the incident response team
B. Increased reporting of security incidents to the incident response team
C. Decreased number of password resets
D. Increased number of identified system vulnerabilities

97. Which of the following approaches is the **BEST** approach to exception management?

A. Escalation processes are defined.
B. Process deviations are not allowed.
C. Decisions are based on business impact.
D. Senior management judgment is required.

98. Which of the following causes the **GREATEST** concern to a risk practitioner reviewing a corporate information security policy that is out of date? The policy:

A. was not reviewed within the last three years.
B. is missing newer technologies/platforms.
C. was not updated to account for new locations.
D. does not enforce control monitoring.

99. Which of the following is **BEST** suited for the review of IT risk analysis results before the results are sent to management for approval and use in decision making?

A. An internal audit review
B. A peer review
C. A compliance review
D. A risk policy review

100. Which of the following will have the **MOST** significant impact on standard information security governance models?

A. Number of employees
B. Cultural differences between physical locations
C. Complexity of the organizational structure
D. Evolving legislative requirements

101. Which of the following **BEST** describes the role of management in implementing a risk management strategy?

A. Ensure that the planning, budgeting and performance of information security components are appropriate.
B. Assess and incorporate the results of the risk management activity into the decision-making process.
C. Identify, evaluate and minimize risk to IT systems that support the mission of the organization.
D. Understand the risk management process so that appropriate training materials and programs can be developed.

102. Senior management has defined the enterprise risk appetite as moderate. A business critical application has been determined to pose a high risk. What is the **BEST** next course of action?

A. Remove the high-risk application and replace it with another system.
B. Request that senior management increase the level of risk they are willing to accept.
C. Determine whether new controls to be implemented on the system will mitigate the high risk.
D. Restrict access to the application to trusted users.

103. Which of the following devices should be placed within a demilitarized zone (DMZ)?

A. An authentication server
B. A mail relay
C. A firewall
D. A router

104. Which of the following is the **MOST** important factor when designing IS controls in a complex environment?

A. Development methodologies
B. Scalability of the solution
C. Technical platform interfaces
D. Stakeholder requirements

105. A **PRIMARY** reason for initiating a policy exception process is when:

A. the risk is justified by the benefit.
B. policy compliance is difficult to enforce.
C. operations are too busy to comply.
D. users may initially be inconvenienced.

106. The IT department wants to use a server for an enterprise database, but the server hardware is not certified by the operating system (OS) or the database vendor. A risk practitioner determines that the use of the database presents:

A. a minimal level of risk.
B. an unknown level of risk.
C. a medium level of risk.
D. a high level of risk.

107. Which of the following approaches to corporate policy **BEST** supports an enterprise's expansion to other regions, where different local laws apply?

A. A global policy that does not contain content that might be disputed at a local level
B. A global policy that is locally amended to comply with local laws
C. A global policy that complies with law at corporate headquarters and that all employees must follow
D. Local policies to accommodate laws within each region

108. Who grants formal authorization for user access to a protected file?

A. The process owner
B. The system administrator
C. The data owner
D. The security manager

109. Which of the following **BEST** addresses the risk of data leakage?

A. Incident response procedures
B. File backup procedures
C. Acceptable use policies (AUPs)
D. Database integrity checks

110. Which of the following would data owners be **PRIMARILY** responsible for?

A. Intrusion detection
B. Antivirus controls
C. User entitlement changes
D. Platform security

111. Which of the following uses risk scenarios when estimating the likelihood and impact of significant risk to the organization?

A. An IT audit
B. A security gap analysis
C. A threat and vulnerability assessment
D. An IT security assessment

112. Which of the following provides the **BEST** view of risk management?

A. An interdisciplinary team
B. A third-party risk assessment service provider
C. The enterprise's IT department
D. The enterprise's internal compliance department

113. Which of the following is the **BEST** risk identification technique for an enterprise that allows employees to identify risk anonymously?

A. The Delphi technique
B. Isolated pilot groups
C. A strengths, weaknesses, opportunities and threats (SWOT) analysis
D. A root cause analysis

114. Information that is no longer required to support the main purpose of the business from an information security perspective should be:

A. analyzed under the retention policy.
B. protected under the information classification policy.
C. analyzed under the backup policy.
D. protected under the business impact analysis (BIA).

115. Which of the following **MOST** effectively ensures that service provider controls are within the guidelines set forth in the organization's information security policy?

A. Service level monitoring
B. Penetration testing
C. Security awareness training
D. Periodic auditing

116. The **MOST** effective method to conduct a risk assessment on an internal system in an organization is to start by understanding the:

A. performance metrics and indicators.
B. policies and standards.
C. recent audit findings and recommendations.
D. system and its subsystems.

117. Investments in risk management technologies should be based on:

A. audit recommendations.
B. vulnerability assessments.
C. business climate.
D. value analysis.

118. Which of the following is **MOST** relevant to include in a cost-benefit analysis of a two-factor authentication system?

A. The approved budget of the project
B. The frequency of incidents
C. The annual loss expectancy (ALE) of incidents
D. The total cost of ownership (TCO)

119. IT risk is measured by its:

A. level of damage to IT systems.
B. impact on business operations.
C. cost of countermeasures.
D. annual loss expectancy (ALE).

120. Which of the following is used to determine whether unauthorized modifications were made to production programs?

A. An analytical review
B. Compliance testing
C. A system log analysis
D. A forensic analysis

121. Which of the following is a **PRIMARY** consideration when developing an IT risk awareness program?

A. Why technology risk is owned by IT
B. How technology risk can impact each attendee's area of business
C. How business process owners can transfer technology risk
D. Why technology risk is more difficult to manage compared to other risk

122. What indicates that an enterprise's risk practices need to be reviewed?

A. The IT department has its own methodology of risk management.
B. Manufacturing assigns its own internal risk management roles.
C. The finance department finds exceptions during its yearly risk review.
D. Sales department risk management procedures were last reviewed 11 months ago.

123. Which of the following **BEST** determines compliance with the risk appetite of an enterprise?

A. Balance between preventive and detective controls
B. Inherent risk and acceptable risk level
C. Residual risk and acceptable risk level
D. Balance between countermeasures and preventive controls

124. Which of the following would **PRIMARILY** help an enterprise select and prioritize risk responses?

A. A cost-benefit analysis of available risk mitigation options
B. The level of acceptable risk per risk appetite
C. The potential to transfer or eliminate the risk
D. The number of controls necessary to reduce the risk

125. In the risk management process, a cost-benefit analysis is **MAINLY** performed:

A. as part of an initial risk assessment.
B. as part of risk response planning.
C. during an information asset valuation.
D. when insurance is calculated for risk transfer.

126. Which of the following is the **PRIMARY** reason for conducting periodic risk assessments?

A. Changes to the asset inventory
B. Changes to the threat and vulnerability profile
C. Changes in asset classification levels
D. Changes in the risk appetite

127. Which of the following factors should be analyzed to help management select an appropriate risk response?

A. The impact on the control environment
B. The likelihood of a given threat
C. The costs and benefits of the controls
D. The severity of the vulnerabilities

128. An enterprise has just completed an information systems audit and a large number of findings have been generated. This list of findings is **BEST** addressed by:

A. a risk mitigation plan.
B. a business impact analysis (BIA).
C. an incident management plan.
D. revisions to information security procedures.

129. A risk assessment indicates a risk to the enterprise that exceeds the risk acceptance level set by senior management. What is the **BEST** way to address this risk?

A. Ensure that the risk is quickly brought within acceptable limits, regardless of cost.
B. Recommend mitigating controls if the cost and/or benefit would justify the controls.
C. Recommend that senior management revise the risk acceptance level.
D. Ensure that risk calculations are performed to revalidate the controls.

130. Malware has been detected that redirects users' computers to web sites crafted specifically for the purpose of fraud. The malware changes domain name system (DNS) server settings, redirecting users to sites under the hackers' control. This scenario **BEST** describes a:

A. man-in-the-middle (MITM) attack.
B. phishing attack.
C. pharming attack.
D. social engineering attack.

131. Which of the following is the **BEST** option to ensure that corrective actions are taken after a risk assessment is performed?

A. Conduct a follow-up review.
B. Interview staff member(s) responsible for implementing the corrective action.
C. Ensure that an organizational executive documents that the corrective action was taken.
D. Run a monthly report and verify that the corrective action was taken.

132. Which of the following is **MOST** important for effective risk management?

A. Assignment of risk owners to identified risk
B. Ensuring compliance with regulatory requirements
C. Integration of risk management into operational processes
D. Implementation of a risk avoidance strategy

133. A global financial institution has decided not to take any further action on a denial-of-service (DoS) vulnerability found by the risk assessment team. The **MOST** likely reason for making this decision is that:

A. the needed countermeasure is too complicated to deploy.
B. there are sufficient safeguards in place to prevent this risk from happening.
C. the likelihood of the risk occurring is unknown.
D. the cost of countermeasure outweighs the value of the asset and potential loss.

134. Which of the following is the **PRIMARY** reason for periodically monitoring key risk indicators (KRIs)?

A. The cost of risk response needs to be minimized.
B. Errors in results of KRIs need to be minimized.
C. The risk profile may have changed.
D. Risk assessment needs to be continually improved.

135. When developing IT-related risk scenarios with a top-down approach, it is **MOST** important to identify the:

A. information system environment.
B. business objectives.
C. hypothetical risk scenarios.
D. external risk scenarios.

136. Which of the following is of **MOST** concern in a review of a virtual private network (VPN) implementation? Computers on the network are located:

A. at the enterprise's remote offices.
B. on the enterprise's internal network.
C. at the backup site.
D. in employees' homes.

137. Which of the following **BEST** helps identify information systems control deficiencies?

A. Gap analysis
B. The current IT risk profile
C. The IT controls framework
D. Countermeasure analysis

138. Which of the following is **MOST** important for measuring the effectiveness of a security awareness program?

A. Increased interest in focus groups on security issues
B. A reduced number of security violation reports
C. A quantitative evaluation to ensure user comprehension
D. An increased number of security violation reports

139. Which of the following is the **MOST** important requirement for setting up an information security infrastructure for a new system?

A. Performing a business impact analysis (BIA)
B. Considering personal devices as part of the security policy
C. Basing the information security infrastructure on a risk assessment
D. Initiating IT security training and familiarization

140. Which of the following **BEST** describes the risk-related roles and responsibilities of an organizational business unit (BU)? The BU management team:

A. owns the mitigation plan for the risk belonging to their BU, while board members are responsible for identifying and assessing risk as well as reporting on that risk to the appropriate support functions.
B. owns the risk and is responsible for identifying, assessing and mitigating risk as well as reporting on that risk to the appropriate support functions and the board of directors.
C. carries out the respective risk-related responsibilities, but ultimate accountability for the day-to-day work of risk management and goal achievement belongs to the board members.
D. is ultimately accountable for the day-to-day work of risk management and goal achievement, and board members own the risk.

141. What role does the risk professional have in regard to the IS control monitoring process? The risk professional:

A. maintains and operates IS controls.
B. approves the policies for IS control monitoring.
C. determines the frequency of control testing by internal audit.
D. assists in planning, reporting and scheduling tests of IS controls.

142. Who is **MOST** likely responsible for data classification?

A. The data user
B. The data owner
C. The data custodian
D. The system administrator

143. Which of the following is **MOST** effective in assessing business risk?

A. A use case analysis
B. A business case analysis
C. Risk scenarios
D. A risk plan

144. Which of the following is **MOST** important when selecting an appropriate risk management methodology?

A. Risk culture
B. Countermeasure analysis
C. Cost-benefit analysis
D. Risk transfer strategy

145. The **PRIMARY** advantage of creating and maintaining a risk register is to:

A. ensure that an inventory of potential risk is maintained.
B. record all risk scenarios considered during the risk identification process.
C. collect similar data on all risk identified within the organization.
D. run reports based on various risk scenarios.

146. The likelihood of an attack being launched against an enterprise is **MOST** dependent on:

A. the skill and motivation of the potential attacker.
B. the frequency that monitoring systems are reviewed.
C. the ability to respond quickly to any incident.
D. the effectiveness of the controls.

147. A risk response report includes recommendations for:

A. acceptance.
B. assessment.
C. evaluation.
D. quantification.

148. The **PRIMARY** reason for developing an enterprise security architecture is to:

A. align security strategies between the functional areas of an enterprise and external entities.
B. build a barrier between the IT systems of an enterprise and the outside world.
C. help with understanding of the enterprise's technologies and the interactions between them.
D. protect the enterprise from external threats and proactively monitor the corporate network.

149. Which of the following metrics is the **MOST** useful in measuring the monitoring of violation logs?

A. Penetration attempts investigated
B. Violation log reports produced
C. Violation log entries
D. Frequency of corrective actions taken

150. What is the **MOST** important reason for periodically testing controls?

A. To meet regulatory requirements
B. To meet due care requirements
C. To ensure that control objectives are met
D. To achieve compliance with standard policy

CRISC™ Review Questions, Answers & Explanations Manual 2015

SAMPLE EXAM ANSWER AND REFERENCE KEY

Exam Question #	Key	Ref. #	Exam Question #	Key	Ref. #	Exam Question #	Key	Ref. #	Exam Question #	Key	Ref. #
1	A	R4-22	39	D	R1-2	77	A	R1-84	115	D	R4-28
2	A	R1-15	40	D	R1-1	78	B	R2-89	116	D	R2-70
3	C	R1-44	41	A	R2-39	79	A	R2-29	117	D	R2-84
4	C	R4-8	42	D	R3-55	80	B	R3-26	118	D	R3-9
5	B	R4-34	43	C	R1-53	81	C	R1-11	119	B	R2-28
6	D	R2-11	44	C	R4-30	82	C	R2-10	120	B	R2-15
7	B	R2-54	45	A	R3-110	83	D	R1-88	121	B	R1-40
8	D	R2-67	46	C	R3-128	84	C	R3-51	122	A	R2-98
9	B	R2-22	47	B	R3-117	85	C	R1-70	123	C	R1-48
10	A	R3-106	48	D	R1-7	86	C	R3-7	124	A	R3-93
11	D	R1-43	49	A	R3-85	87	D	R2-72	125	B	R3-10
12	B	R3-30	50	B	R4-7	88	B	R4-41	126	B	R4-37
13	B	R4-70	51	B	R1-61	89	D	R3-84	127	C	R2-88
14	C	R4-73	52	D	R2-93	90	B	R1-66	128	A	R4-49
15	C	R2-23	53	B	R2-71	91	A	R4-18	129	B	R3-95
16	C	R4-55	54	B	R1-8	92	B	R2-46	130	C	R1-6
17	A	R1-29	55	D	R1-83	93	C	R4-44	131	A	R3-59
18	B	R1-56	56	A	R2-47	94	C	R2-40	132	A	R1-52
19	C	R2-82	57	C	R1-60	95	A	R1-45	133	D	R3-8
20	C	R4-26	58	D	R2-30	96	B	R1-21	134	C	R4-17
21	D	R1-34	59	C	R2-16	97	A	R2-87	135	B	R1-71
22	D	R4-46	60	D	R2-55	98	A	R4-35	136	D	R2-92
23	B	R2-48	61	C	R2-9	99	B	R2-35	137	A	R2-25
24	B	R1-74	62	A	R3-129	100	C	R2-2	138	D	R4-25
25	C	R3-94	63	D	R1-59	101	B	R2-68	139	C	R1-12
26	C	R4-19	64	B	R3-114	102	C	R2-85	140	B	R2-7
27	D	R1-55	65	A	R4-16	103	B	R3-36	141	D	R4-71
28	A	R4-5	66	A	R4-69	104	D	R3-31	142	B	R2-75
29	B	R3-49	67	C	R3-76	105	A	R3-14	143	C	R1-32
30	D	R4-40	68	A	R4-4	106	B	R2-45	144	A	R1-47
31	C	R3-13	69	D	R1-4	107	B	R1-20	145	A	R1-31
32	D	R4-54	70	B	R1-41	108	C	R3-104	146	A	R1-81
33	A	R1-42	71	B	R2-4	109	C	R3-35	147	A	R3-6
34	C	R2-8	72	D	R3-123	110	C	R3-124	148	A	R4-31
35	A	R1-24	73	C	R2-13	111	C	R2-1	149	A	R4-24
36	A	R3-74	74	C	R4-36	112	A	R1-19	150	C	R4-72
37	A	R4-33	75	C	R3-73	113	A	R1-26			
38	C	R4-29	76	D	R3-105	114	A	R1-3			

Reference example: RS4-9 = See domain 4, question 9, for explanation of the answer.

Page intentionally left blank

CRISC™ Review Questions, Answers & Explanations Manual 2015

SAMPLE EXAM ANSWER SHEET (PRETEST)

(side 1)

Please use this answer sheet to take the sample exam as a pretest to determine strengths and weaknesses. The answer key/reference grid is on page 203.

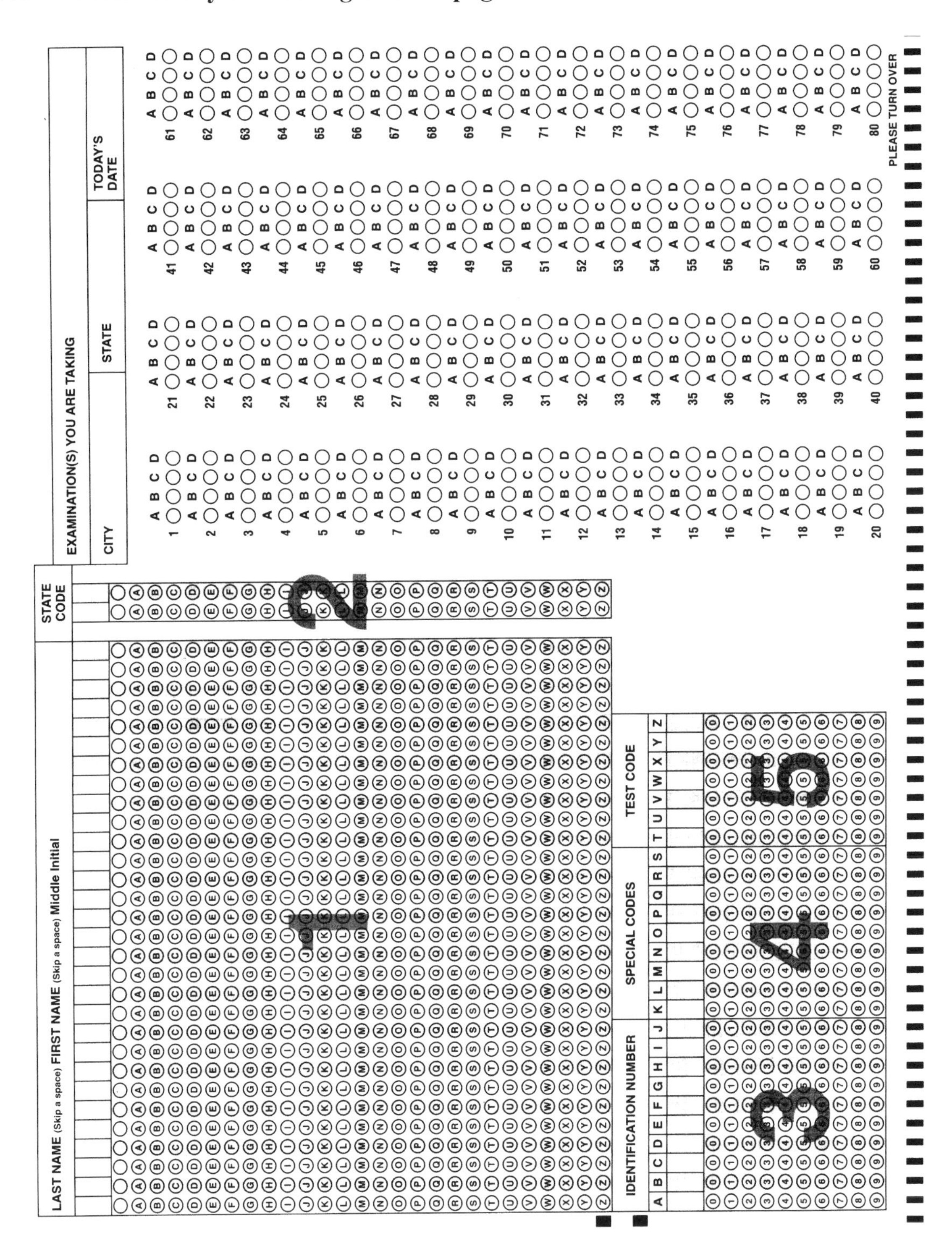

(side 2)

Please use this answer sheet to take the sample exam as a pretest to determine strengths and weaknesses. The answer key/reference grid is on page 203.

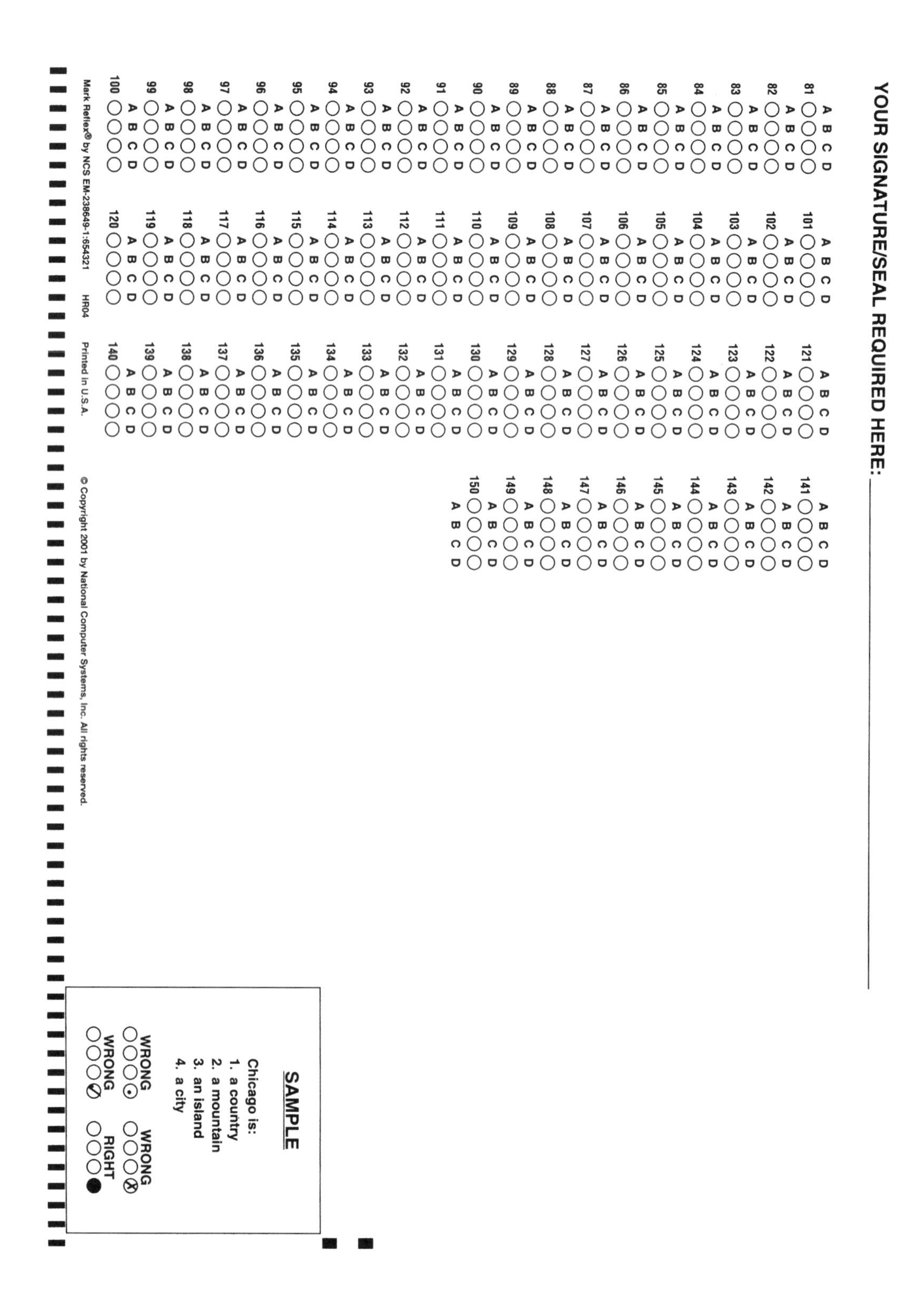

YOUR SIGNATURE/SEAL REQUIRED HERE:

81 A B C D	101 A B C D	121 A B C D	141 A B C D
82 A B C D	102 A B C D	122 A B C D	142 A B C D
83 A B C D	103 A B C D	123 A B C D	143 A B C D
84 A B C D	104 A B C D	124 A B C D	144 A B C D
85 A B C D	105 A B C D	125 A B C D	145 A B C D
86 A B C D	106 A B C D	126 A B C D	146 A B C D
87 A B C D	107 A B C D	127 A B C D	147 A B C D
88 A B C D	108 A B C D	128 A B C D	148 A B C D
89 A B C D	109 A B C D	129 A B C D	149 A B C D
90 A B C D	110 A B C D	130 A B C D	150 A B C D
91 A B C D	111 A B C D	131 A B C D	
92 A B C D	112 A B C D	132 A B C D	
93 A B C D	113 A B C D	133 A B C D	
94 A B C D	114 A B C D	134 A B C D	
95 A B C D	115 A B C D	135 A B C D	
96 A B C D	116 A B C D	136 A B C D	
97 A B C D	117 A B C D	137 A B C D	
98 A B C D	118 A B C D	138 A B C D	
99 A B C D	119 A B C D	139 A B C D	
100 A B C D	120 A B C D	140 A B C D	

SAMPLE

Chicago is:

1. a country
2. a mountain
3. an island
4. a city

WRONG WRONG WRONG RIGHT

Mark Reflex® by NCS EM-238649-1:654321 HR04 Printed in U.S.A.

CRISC™ Review Questions, Answers & Explanations Manual 2015

SAMPLE EXAM ANSWER SHEET (POSTTEST)

(side 1)

Please use this answer sheet to take the sample exam as a posttest to determine strengths and weaknesses. The answer key/reference grid is on page 203.

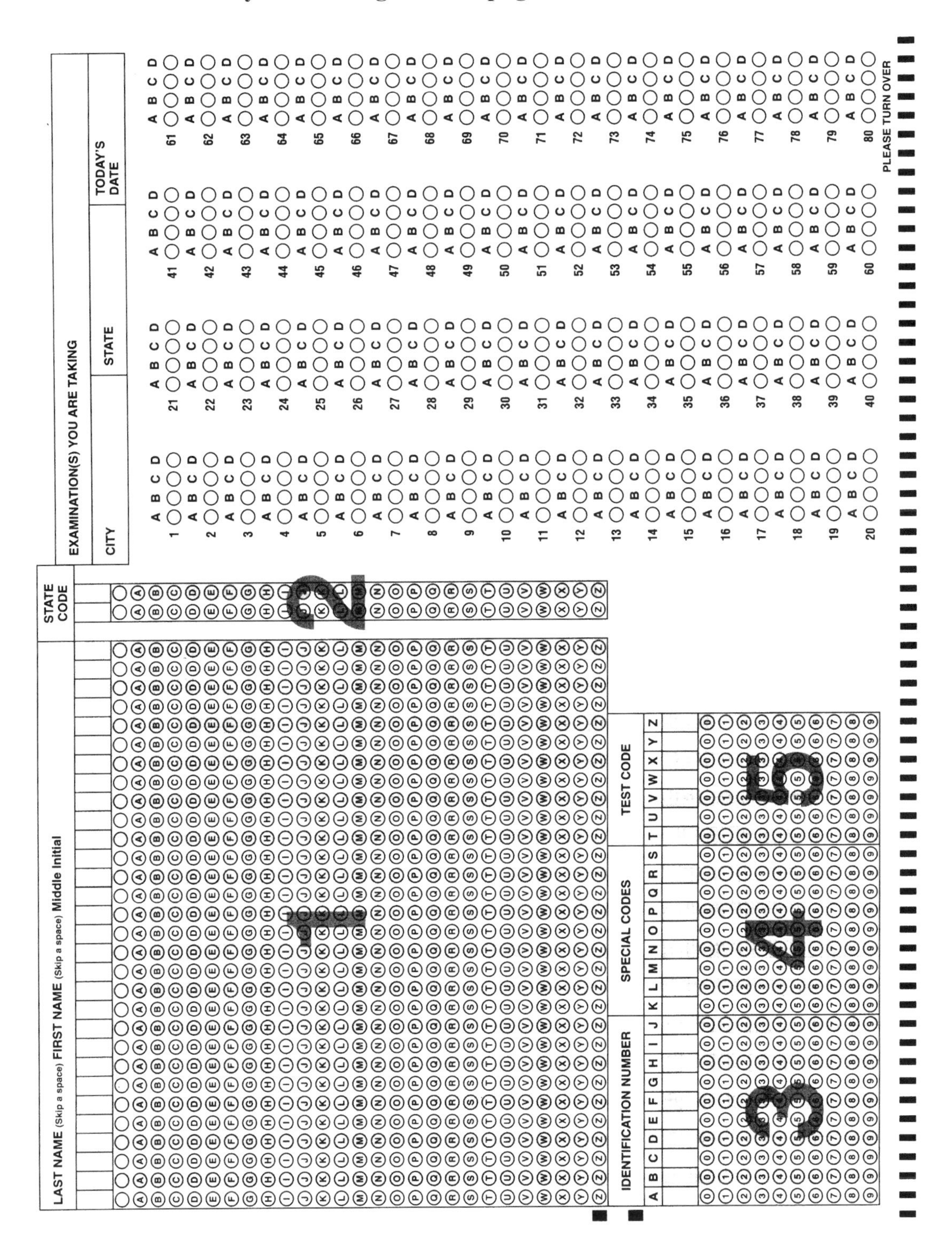

(side 2)

Please use this answer sheet to take the sample exam as a posttest to determine strengths and weaknesses. The answer key/reference grid is on page 203.

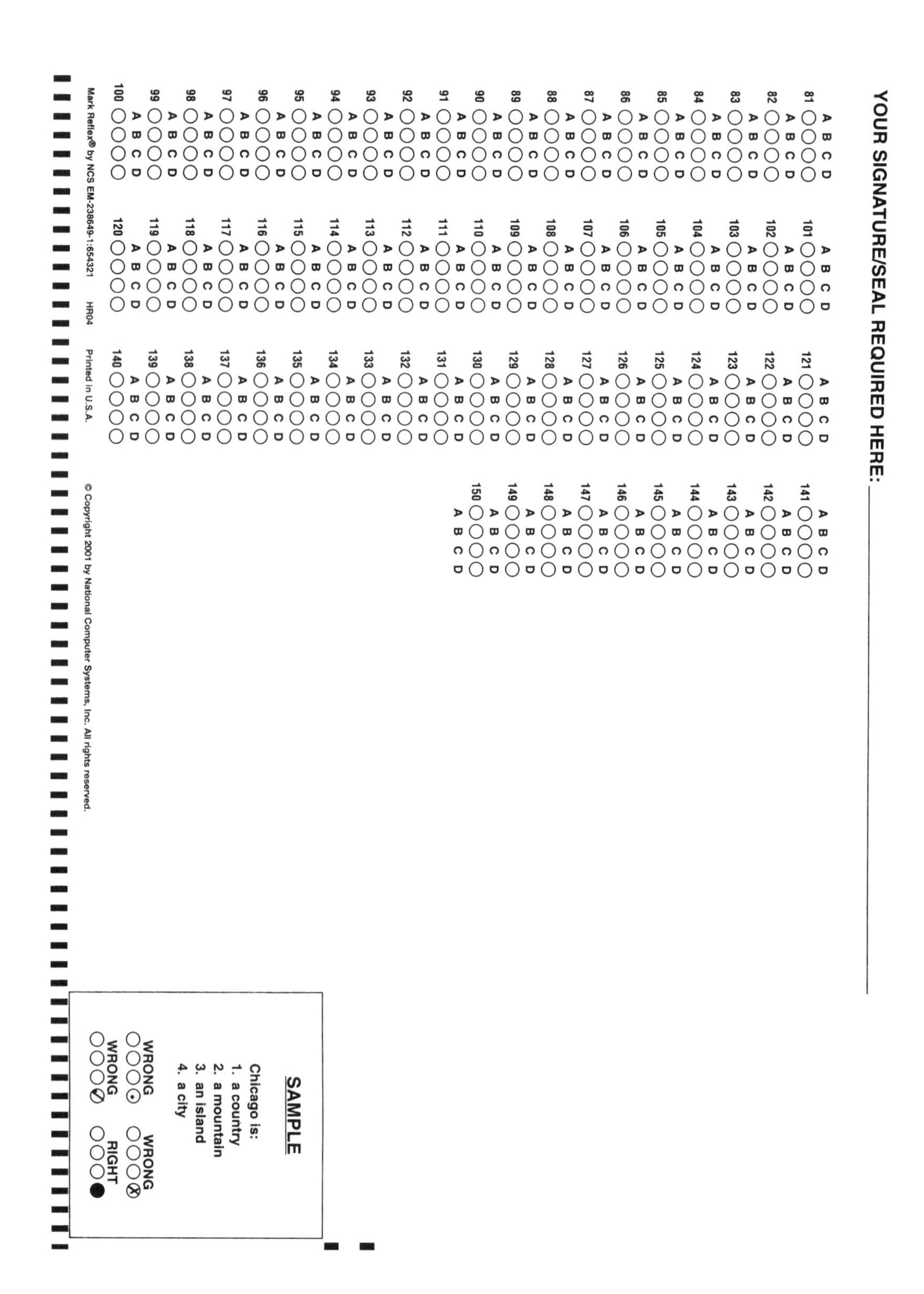

EVALUATION

ISACA continuously monitors the swift and profound professional, technological and environmental advances affecting risk practitioners. Recognizing these rapid advances, CRISC manuals are updated annually.

To assist ISACA in keeping abreast of these advances, please take a moment to evaluate the *CRISC™ Review Questions, Answers & Explanations Manual 2015*. Such feedback is valuable to fully serve the profession and future CRISC exam registrants.

To complete the evaluation on the web site, please go to *www.isaca.org/studyaidsevaluation.*

Thank you for your support and assistance.